The Land Is Our Community

The Land Is Our Community

ALDO LEOPOLD'S
ENVIRONMENTAL ETHIC FOR
THE NEW MILLENNIUM

Roberta L. Millstein

THE UNIVERSITY OF CHICAGO PRESS
CHICAGO AND LONDON

This book is freely available in an open access digital edition thanks to TOME (Toward an Open Monograph Ecosystem)—a collaboration of the Association of American Universities, the Association of University Presses, and the Association of Research Libraries—and the generous support of the University of California, Davis. Learn more at the TOME website: openmonographs.org.

The terms of the license for the open access digital edition are Creative Commons Attribution-Non-Commercial-No-Derivatives 4.0 International License (CC BY-NC-ND 4.0). To view a copy of this license, visit https://creativecommons.org/licenses/by-nc-nd/4.0/.

The University of Chicago Press, Chicago 60637
The University of Chicago Press, Ltd., London
© 2024 by The University of Chicago
Subject to the exception mentioned above, no part of this book may be used or reproduced in any manner whatsoever without written permission, except in the case of brief quotations in critical articles and reviews. For more information, contact the University of Chicago Press, 1427 E. 60th St., Chicago, IL 60637.
Published 2024
Printed in the United States of America

33 32 31 30 29 28 27 26 25 24   1 2 3 4 5

ISBN-13: 978-0-226-83446-7 (cloth)
ISBN-13: 978-0-226-83448-1 (paper)
ISBN-13: 978-0-226-83447-4 (e-book)
DOI: https://doi.org/10.7208/chicago/9780226834474.001.0001

Library of Congress Cataloging-in-Publication Data

Names: Millstein, Roberta L., author.
Title: The land is our community : Aldo Leopold's environmental ethic for the new millennium / Roberta L. Millstein.
Other titles: Aldo Leopold's environmental ethic for the new millennium
Description: Chicago ; London : The University of Chicago Press, 2024. | Includes bibliographical references and index.
Identifiers: LCCN 2023050820 | ISBN 9780226834467 (cloth) | ISBN 9780226834481 (paperback) | ISBN 9780226834474 (ebook)
Subjects: LCSH: Leopold, Aldo, 1886–1948. | Environmental ethics. | Ecology—Moral and ethical aspects. | Ecology—Philosophy. | Nature conservation—Philosophy.
Classification: LCC QH31.L618 M55 2024 | DDC 179/.1—dc23/eng/20231212
LC record available at https://lccn.loc.gov/2023050820

♾ This paper meets the requirements of ANSI/NISO Z39.48-1992 (Permanence of Paper).

To all my fellow-voyagers in the odyssey of evolution

It is interesting to contemplate an entangled bank, clothed with many plants of many kinds, with birds singing on the bushes, with various insects flitting about, and with worms crawling through the damp earth, and to reflect that these elaborately constructed forms, so different from each other, and dependent on each other in so complex a manner, have all been produced by laws acting around us.

CHARLES DARWIN, *On the Origin of Species*

Contents

Preface · xi

CHAPTER ONE
Reinterpreting Leopold · 1

CHAPTER TWO
Interdependence · 27

CHAPTER THREE
Land Communities · 51

CHAPTER FOUR
Land Health · 79

CHAPTER FIVE
Arguing for the Land Ethic · 111

CHAPTER SIX
Policy Implications · 135

References · 161
Index · 175

Preface

I came to Aldo Leopold's land ethic in an unusual way. Most people, I have come to realize, read *A Sand County Almanac* and find themselves captivated by Leopold's poetic turns of phrase, his ability to bring to life the natural world around us and make it personal, his powerful and hard-hitting prose in essays like "Thinking like a Mountain" and "On a Monument to the Pigeon." Being of a philosophical bent, I came to the land ethic via a much more analytical, and also circuitous, route. For many years his essay "The Land Ethic" was simply one of many readings I would assign in my environmental ethics classes—plausible, to be sure, but in my mind just one possible way of approaching the topic among a number of other candidates.

There is no one single moment when my mind changed, but I can think of several "aha" moments over a number of years, all involving teaching, that pointed me to Leopold. One was the day my students asked me yet again why one of our authors was interpreting Leopold in the way that they were, and once again I had to say that I didn't see any justification for that interpretation. But I had finally had enough. From that day forward, I vowed that in my classes, we would read Leopold's own words for ourselves and find our own interpretations. Then there was the very fortuitous day when we were reading "The Land Ethic" in one of my classes and chapter 3 of Charles Darwin's *On the Origin of Species* in another, and I realized that Leopold was talking about what Darwin was talking about— the interdependence of species in the struggle for existence. That led me to thinking about a whole new way of understanding the land ethic, with the interdependence between species—including humans—at the center.

Other significant days included the day that I recognized that the most plausible way to read Leopold was as someone who valued not just the land communities that we are interdependent with (a holistic way of thinking), but also all the individual *members* of the land community and all the ways that they could be valuable, including aesthetically. Then there was the day I realized that, whereas my students (who were from mixed academic backgrounds) often fought tooth and nail against animal rights perspectives, seeing them as too draconian despite my best efforts to make them sound credible, when we got to the land ethic, they were by and large on board. For students who had backgrounds in environmental studies, the affinity for Leopold's way of thinking was even clearer. At some point I just started to think that an environmental ethic with such widespread appeal was worth a deeper look. (There was also the day that the Leopold documentary *Green Fire* was screened on the University of California, Davis, campus; people arrived in droves clutching dog-eared and bookmarked copies of *A Sand County Almanac*.)

In short, I began to see that a new interpretation and new defense of Leopold was sorely needed and likely to be appreciated. My first published paper on the land ethic delved into the issue of understanding Leopold in light of Darwin's ecological views (yes, ecological) on the interdependence of species. And once I did that, I realized that there were a number of central concepts that required further explication too: land community, land health, and interdependence itself. As a philosopher of biology trained to analyze central concepts in science through a historical lens, I thought I could make a useful contribution there. But as my published papers began to pile up, I realized that I was really dealing with issues that ought to be together in a book, not as separate papers. Thus, the idea for this book was born.

Thinking about Leopold also made me think about one of the things that attracted people to his views—the fact that he was not born to the land ethic, but instead came to it over time after a lifetime of experiences. That made him human—one of us. One of those experiences is documented in a fictionalized form in "Thinking like a Mountain," where Leopold (who was a hunter) recounts killing a wolf and feeling regret as he saw the green fire in her eyes die. That made me think about what story I would tell of how I came to care about environmental issues. I am truly not sure, and even if I were, I could not possibly express it as well as Leopold did, but one event sticks in my mind. As a child growing up in New Jersey, I often spent part of the summer hiking in the White Mountains of New Hampshire with my family. Even from a young age, I loved gazing out at the soft, rounded mountains that seemed to go on forever. But vacations must end, and as we were driving home one time—perhaps I was around ten years old—I was looking at the industrial smokestacks along the New Jersey Turnpike, and it struck me that *we* had done this: we had cut down all the trees and replaced them with miles and miles of asphalt and cement. And I started to cry. If I am honest, it still makes me cry to think about it.

Many other experiences have influenced my views concerning the environment, but some particularly relevant ones took place much later here in Davis, California. For a little over ten years, I served on the city's Open Space and Habitat Commission, thinking it was important to try to put my more theoretical ideas into practice. I learned the value of the surrounding farmlands—many Davisites take advantage of these at our biweekly farmers market or through weekly Community Supported Agriculture (CSA) vegetable and fruit boxes—as well as the subtle habitats in and around the city. Perhaps the most important thing I learned, though, were the challenges, successes, and failures involved in working with a group of people to try to bring ideas to the local city council for approval. I learned that it is in fact possible to come to consensus if everyone is committed to making things work, and if at least some goals and values are shared, but also how sensitive such balances can be. I couldn't help but read about Leopold's own commission work in light of my experiences and feel his successes and failures along with him. Putting a land ethic into practice would never be easy, but it could be highly rewarding.

Additionally, Leopold's emphasis on *community* very much resonated with me—not just because of my local community, but also because of the members of the academic community who surround and support me. Indeed, that community has been with me every step of the way, and so I must give thanks to many people at various venues (conferences, department colloquia, social media, etc.) who heard me present ideas that made their way into this book, who asked me hard questions and gave me thoughtful suggestions. The philosophy of biology lab group that I co-run with Jim Griesemer at UC Davis read many drafts and offered constructive feedback over the course of my years writing about and thinking about Leopold, a changing cast of people who never failed to make my work better. More explicitly, I would like to thank the following people for helpful comments on various parts and stages of this book, including previously published work: Marshall Abrams, Holly Andersen, Antoine C. Dussault, Chris Eliot, Alkistis Elliott-Graves, Justin Garson, Jim Griesemer, Dan Hicks, Denise Hossom, Andrew Inkpen, Connor Kianpour, Kip Koelsch, Chris Lean, Stephen Linquist, Clement Loo, Katie McShane, Curt Meine, Rick Morris, Jonathan Newman, Jay Odenbaugh, Maureen O'Malley, Steve Peck, Anya Plutynski, Sarah Roe, Tami Schneider, Gary Varner, Kenneth Blake Vernon, Julianne Warren, Chris Young, plus numerous anonymous referees. Jeff Ramsey deserves special thanks for providing my first introduction to environmental ethics. I am also grateful to the University of Chicago Press—particularly Executive Editor Karen Darling and Editorial Associate Fabiola Enríquez—for all their assistance and willingness to work with me to figure out how to publish this book open access, not to mention UC Davis for providing the funding to publish open access in the first place.

Last but not least, I would be remiss if I didn't thank my partner Gilbert, who has patiently listened to me bounce half-formed thoughts off him when all else was failing. I am deeply indebted to my parents, Norman and Rita Millstein, for taking my sister and me on many hikes, helping to instill in us a love for all things wild. I also need to thank all the nonhuman members of my community, including but not limited to the trees, the flowers and the grass, the squirrels and birds, the corn and the tomatoes, the worms, the soil, and, especially, my poodles.

CHAPTER ONE

Reinterpreting Leopold

No important change in ethics was ever accomplished without an internal change in our intellectual emphasis, loyalties, affections, and convictions. The proof that conservation has not yet touched these foundations of conduct lies in the fact that philosophy and religion have not yet heard of it. In our attempt to make conservation easy, we have made it trivial.

ALDO LEOPOLD, *A Sand County Almanac*

Parts of this chapter originally appeared in Roberta L. Millstein, "Debunking Myths about Aldo Leopold's Land Ethic," *Biological Conservation* 217 (2018): 391–96.

INTRODUCTION

This is a book about Aldo Leopold's land ethic,[1] a view he developed over the course of his lifetime, a view that was informed by his experiences as a hunter, forester, wildlife manager, ecologist, conservationist, and professor. It culminated in the essay "The Land Ethic" in *A Sand County Almanac*, published posthumously after his untimely death at age sixty-one in 1948. It has been extremely influential in environmental ethics as well as conservation biology and related fields, especially the fields that he was involved with. The land ethic called for an expansion of our ethical obligations beyond the purely human to include what he variously called the "land community" or the "biotic community"—communities of interdependent humans, nonhuman animals, plants, soils, and waters, understood collectively.

Using an approach grounded in environmental ethics and the history and philosophy of science, I offer a new interpretation of Leopold's land ethic and a new defense of it in light of contemporary ecology. Despite the enormous influence of the land ethic, it has sometimes been prematurely dismissed as either empirically out of date or ethically flawed. However, these dismissals are unfounded; they are based on problematic interpretations of Leopold's land ethic. Previous interpretations have failed to do a proper historical and philosophical (conceptual) analysis; they have failed to provide

---

1. To be clear, Leopold spoke of the land ethic as a continually evolving product of social evolution; I focus primarily on the version that he had developed by the time of his death rather than other versions that have been developed since or could be developed in the future.

sufficient textual evidence and have failed to take into account relevant parts of Leopold's life and work.[2] In this book, I provide new, more defensible interpretations of the central concepts underlying Leopold's land ethic: interdependence, land community, and land health. I also provide a new and more defensible interpretation of his argument for extending our ethics to include land communities and Leopold-inspired guidelines for how the land ethic can guide conservation and restoration policy.

On my interpretation, the argument for the land ethic is actually quite simple: if we accept ethical rules (limitations on our actions for dealing with individuals and society) because of our interdependence with other humans, then once we recognize that we are interdependent with other species, soils, and waters, we ought to extend our ethics to include our land communities as well. But beneath the apparent simplicity of the argument lies complexity, because there is much to spell out here in terms what a land community is, what interdependence in a land community is, what it means for a land community to be healthy, why we should think that a land community has value in its own right, and what the policy implications of the land ethic are. Spelling all of these ideas out, which will result in a defense of the land ethic, is the objective of this book.

In this chapter, I start with a brief overview of Aldo Leopold's life and his influence. I then make the case for why Leopold should be reinterpreted, including a discussion of the ways in which some existing interpretations are problematic and should be rejected. I then provide an overview of the book, including short chapter summaries.

ALDO LEOPOLD AND HIS INFLUENCE

I begin with an extremely brief outline of some significant events in Aldo Leopold's life. For outstanding biographies of Leopold, see Flader (1994), Meine (2010), and Warren (2016). These histories of Leopold's life and research are highly recommended and have been

---

2. According to the online Aldo Leopold Archives, he published more than five hundred articles, essays, and reports. The archives contain many of his unpublished essays as well, some of which have since been published in edited volumes.

invaluable to me in the writing of this book. The summary here relies on the "Chronology" in *A Sand County Almanac & Other Writings on Conservation and Ecology* (Meine 2013).

Aldo was born in Burlington, Iowa, on January 11, 1887. He was the oldest of four children of Carl and Clara Starker Leopold. His mother was the daughter of a prominent Burlington businessman and civic leader; his father was a co-partner of what would become the Leopold Desk Company. From an early age, he engaged in a number of outdoor activities, eventually to include camping, sailing, fishing, and bird-watching. When Aldo was twelve, his father began teaching him how to hunt, including his views on how to hunt ethically. Family trips were outdoor-oriented, and destinations included Michigan, Illinois, and Colorado.

Leopold began attending Yale University in 1905, where he learned forestry, earning a BS and an MS. In 1909, he was appointed Forest Assistant on the Apache National Forest in Arizona. Not long after starting, he was appointed the head of a forest reconnaissance crew whose job it was to map the land and survey the timber, but he was inexperienced, and it was a bit rocky. He met his wife-to-be, Estella, in 1911 while on a work trip to Albuquerque, New Mexico; they were married in 1912, not long after he had been promoted to Forest Supervisor. In 1913, he almost died from acute nephritis and had to spend the next eighteen months convalescing. When he returned to work in September 1914, he was assigned to the office of grazing in District 3 in Albuquerque, and his family relocated to Albuquerque. In 1915 he was given responsibility for oversight of US Forest Service (USFS) work in his district on recreation, publicity, and game and fish conservation. He made several trips to the Grand Canyon to prepare a working plan for the management of the area, began to organize "game protective associations" throughout the Southwest, and was otherwise active in promoting game conservation.

In 1919, he became Assistant District Forester in Charge of Operations for twenty million acres of USFS land in the Southwest, where he had the opportunity to study "the interrelationship of historic and contemporary grazing, soil erosion, vegetation change, and climate in the ecological functioning of southwestern watersheds" (Meine 2013, 852). In 1922, he recommended the designation of a 755,000-acre portion of the Gila National Forest in New

Mexico as a wilderness area, which was approved in 1924. During this time, he began to pen his thoughts about conservation and human responsibilities toward the land, and he accepted a transfer to Madison, Wisconsin, to become the Assistant (later Associate) Director of the USFS Forest Products Laboratory. He started to publish on the topics of wilderness values, wilderness protection, and game management. In 1928, he resigned his position with the Forest Products Laboratory and left the USFS to conduct statewide game surveys as a private consultant, eventually including surveys of Michigan, Minnesota, Iowa, Ohio, Mississippi, Illinois, Indiana, Wisconsin, and Missouri.

In 1932 funding was withdrawn for his position, and he was temporarily without work, but in 1933 he found short-term employment with the USFS in the spring and summer, overseeing erosion-control projects of the new Civilian Conservation Corps in the Southwest. In July, he accepted an appointment as the new chair of game management in the Department of Agricultural Economics at the University of Wisconsin; he shortly began teaching and accepted graduate students. In 1935, he purchased an eighty-acre farm in rural Wisconsin, on which he renovated an abandoned chicken coop that would become his family's "Shack." During the same year, he visited Germany and Czechoslovakia, touring state farms and land estates. In 1936, Aldo and his family (by this point he and Estella had five children) began a restoration effort on the Shack property, starting with the planting of two thousand pine trees and other trees and shrubs.

Throughout his subsequent years as a University of Wisconsin professor, teaching and supervising students, he also wrote essays that would later become parts of, or revised parts of, *A Sand County Almanac* (ASCA). In 1943, he was appointed to the Wisconsin Conservation Commission, which grappled with (among other issues) deer management. In 1947, he was elected honorary vice president of the American Forestry Association and president of the Ecological Society of America, and he submitted the manuscript that would become ASCA (originally entitled "Great Possessions") to Oxford University Press. The book was accepted for publication on April 14, 1948. He died from a heart attack on April 21, 1948 while helping to fight a fire on a farm near the Shack. ASCA was published posthumously in 1949. It has been translated into fifteen languages.

This brief summary of some of the key events in Leopold's life (again, this is no substitute for existing biographies) is meant to show the diversity of hats that Leopold wore together with the diversity of his experiences (and if this were more complete, it would also include things he read, people he met, places that he traveled to on vacation, and numerous publications and presentations). These hands-on experiences form the backdrop and the basis for his land ethic; I will refer to many of them explicitly or implicitly throughout the rest of this book. Especially notable are his experiences with hunting and the outdoors in many locations, with forestry and trying to manage forests, with erosion and grazing, with game management (especially deer), with his restoration efforts at the Shack, with his trips to Germany and the Sierra Madre (the latter in 1936 and 1937), and with the Civilian Conservation Corps and the Wisconsin Conservation Commission. I mention each of these at various points throughout this book.

As noted above, Leopold's ideas have been extremely influential. Philosopher Eric Katz writes, "Leopold's classic essay 'The Land Ethic' in *A Sand County Almanac* is probably the most widely cited source in the literature of environmental philosophy. His view of the moral consideration of the land-community is the starting point for almost all discussions of environmental ethics" (Katz 1996, 113). Indeed, it would be an unusual environmental ethics course of any breadth that did not discuss Leopold's land ethic, and there have been many publications supporting, challenging, and elaborating on his ideas in the environmental ethics literature.

Similarly, biologist/environmentalist Fred Van Dyke writes,

> Leopold's original contribution was to combine this ethical conservation with practical experience in resource management, and then to inform both with scientific expertise.... [He] began to change fundamental assumptions not only about the best use of natural resources but also about the nature and purpose of ecological studies. These changes opened the door for the development of a value-driven approach to science and conservation, *without which the field of conservation biology could not have emerged*.... Today many conservation biologists see themselves as heirs of Leopold's legacy to restore ethics and value to the science of conservation. (Van Dyke 2008, 41; emphasis added)

The degree of Leopold's influence is perhaps not surprising. His writings melded his scientific knowledge, his hands-on practical experience, his breadth of expertise across conservation subfields, and his respect for the natural world. In ASCA as a whole and in the essay "The Land Ethic" in particular, he sought to inspire not only action but reflection, recognizing that values drive actions and that facts alone would not be sufficient for conservation. But the book did not come out of nowhere. ASCA came from a lifetime of his own reflections that resulted in hundreds of written works produced for a variety of audiences: scientific, practical, and political. His lifetime of reflecting on these values informed his science, and his science informed his values, producing groundbreaking results in both and anticipating many issues that remain live today.[3]

WHY REINTERPRET LEOPOLD?

Revisiting the ideas of a twentieth-century ecologist and conservationist might seem like an odd thing to do; one might think that his ideas would have to be sorely out of date. However, as I show throughout the book, this simply is not the case. Leopold often anticipated issues that we are still grappling with today, such as the central importance of soil fertility and biodiversity (see chapter 4).

Yet, with so many Indigenous environmentalist scholars, in particular, finally receiving the broader attention that they deserve—with much more work still to do to give ear to these voices—a white, male ecologist from the last century might seem like an inappropriate choice for a book focus. These are delicate and complicated issues, and I cannot hope to fully address them here. But here are some thoughts. There is the suggestion that Leopold may have been influenced by Indigenous views (Shilling 2009), implying the possibility of important connections that could be drawn. For example, Kimmerer's (2013) discussion of the importance of seeing reciprocity and familial relations between the human and nonhuman might be echoed in some form in Leopold as interdependence

---

3. Much recent work in the philosophy of science examines what the connections between science and values are and ought to be. Leopold is potentially a very instructive figure in this regard, a topic that I hope to examine in a future work.

and kinship, respectively.[4] Whyte (2015) acknowledges the possibility of drawing these types of connections, but he cautions against purely abstract comparisons like this that overlook considerable differences between Indigenous approaches and those of colonial settlers. Both the potential for comparisons and the need for caution if one makes them are important to keep in mind.

Curt Meine examines another strand of this issue, arguing that "[c]onclusions about Leopold's attitudes on race, social justice, and social progress should take into account the totality of his life experience, acknowledging his faults as well as his evolving vision" (Meine 2022, 168). Although the conversation that Meine productively outlines is not the focus of this book, I hope that the book can contribute indirectly by at least offering a more accurate interpretation of Leopold than what has come before. More generally, since I am not myself indigenous to North America, I am not in a position to properly convey such ideas. On the other hand, as a philosopher of biology steeped in history of science, environmental ethics, and ecology, I am in a position to convey and defend the ideas of an influential thinker to those working in the areas that he has already had an impact on (again, environmental ethics, conservation biology, forestry, etc.).

For Leopold, human issues, which include social/equity issues, cannot be disentangled from nonhuman ones. To think that we can focus only on one or the other is a false and dangerous choice. If there is any reason to think that Leopold had profound insights about ethics and the natural world that are still important today, those insights should be understood correctly. There may be more for Leopold to teach us; indeed, one claim of this book is that there is more, and that what Leopold actually was trying to teach us is more defensible and more consistent with contemporary science than what some have thought he was trying to teach us. The result

---

4. With respect to kinship, Leopold wrote, "It is a century now since Darwin gave us the first glimpse of the origin of species. We know now what was unknown to all the preceding caravan of generations: that men are only fellow-voyagers with other creatures in the odyssey of evolution. This new knowledge should have given us, by this time, *a sense of kinship with fellow-creatures*; a wish to live and let live; a sense of wonder over the magnitude and duration of the biotic enterprise" (1949, 109; emphasis added).

is an ethical basis for our conservation policies that is more well-informed and defensible.

There are several reasons why a revisioning of Leopold's land ethic is needed. One is simply to set the record straight on a central figure in the conservation movement of his day, a person who would ultimately become a central figure in the subsequent environmental movements of the 1960s and 1970s through today. More importantly, Leopold is potentially a unifying figure, given his existing influence across so many different domains—domains that have not always communicated well with one another. Arguably, he has been influential across disciplines because Leopold's land ethic is grounded in a lifetime of practical, hands-on learning—mistakes and all—about the environment and environmental values. As a result, Leopold's land ethic, when properly interpreted, is a practical and appealing view for how we ought to live in the world. By including humans and human practices, it gives guidance for farming, forestry, restoration, and other human interactions with the environment, but it does so while also recognizing and taking into account the value of the organisms and land communities that are impacted by our actions (i.e., while recognizing that their value is more than just their usefulness to us). It is a balanced and comprehensive approach. And its message is all too timely, given the recent international warnings concerning climate change and biodiversity depletion. We desperately need an ethical basis for our actions that is persuasive, motivational, practical, and truly (literally) global.

We are experiencing a multifaceted, global environmental crisis, one that is almost entirely (or entirely, period) the result of human actions. To address it, we want all good ideas on the table for consideration. As discussed in the next section, some authors state that they reject Leopold's views, but those rejections are based on misunderstandings and not Leopold's actual views. The rejections are thus hasty. On the other hand, the picture of the land ethic that emerges after debunking those misunderstandings is one that is appealing and practical. A recent debate has been described as "The Battle for the Soul of Conservation Science" (Kloor 2015), a debate that contrasts the traditional view in conservation biology as preservationist (often associated with Leopold) with one in which humans play a more active and even constructionist role. The view of Leopold presented here will show that there is another alternative to these two extremes. Relatedly, the perceived need for prioritizing ecosystems (again, a view associated with Leopold) is sufficiently high as to have spawned a new journal, *The Ecocentric Citizen*. An opinion piece coauthored by editors of the journal characterizes ecocentrism as a view that holds that "human needs, like the needs of other species, are secondary to those of the Earth as the sum of its ecosystems" (Gray et al. 2017). But was this Leopold's view, as some have suggested, and are there other plausible alternatives? Debunking the myths surrounding Leopold will reveal another path—one that is sympathetic to ecocentrism as defined by Gray et al. in some respects but finds a middle ground.

DEBUNKING MYTHS ABOUT LEOPOLD'S LAND ETHIC

A number of misunderstandings have grown up around Leopold's land ethic and have become so entrenched as to have the status of myths.[5] I identify six such myths:

Myth 1: There is a two-sentence "summary moral maxim" of the land ethic.
Myth 2: When Leopold said "biotic community," he meant "ecosystem."
Myth 3: Ecosystems are the only entities of value in the land ethic.
Myth 4: The central message of the land ethic is to set aside human-free ecosystems.
Myth 5: By "stability," Leopold meant something like "balance" or "dynamic equilibrium."
Myth 6: Leopold's ethics are derived from Charles Darwin's "proto-sociobiological" perspective on ethical phenomena.

Each myth is described in further detail below. The goal of this section is to debunk each of these myths to clear space for the alternative interpretations I provide in the rest of the book.

---

5. The use of the term myth is meant only to indicate the widespread persistence of these mistaken beliefs over time and their transmission from person to person; the term has other connotations and associations (such as an association with traditional cultures), but those connotations and associations should not be inferred by the reader in this instance.

Myth 1: There is a two-sentence "summary moral maxim" of the land ethic.

It is claimed that the following quote from Leopold's essay "The Land Ethic" is the "summary moral maxim" of the land ethic:[6]

> A thing is right when it tends to preserve the integrity, stability, and beauty of the biotic community. It is wrong when it tends otherwise. (Leopold 1949, 224–25)

The implication is that the essence of the land ethic can be gleaned from these two sentences. Even without the phrase "summary moral maxim," these two sentences are often treated as a summary of Leopold's land ethic. For example, Tom Regan quotes these two sentences and from them alone infers that the "implications of this view include the clear prospect that the individual may be sacrificed for the greater biotic good" (Regan 1983, 361). Having made that quick inference, Regan just as quickly rejects the land ethic for endorsing "environmental fascism."

Similarly, Ned Hettinger and Bill Throop (1999) quote the same two sentences as a "summary maxim" of the land ethic and then proceed directly to a criticism of Leopold's use of the term *stability*. They equate stability with equilibrium and balance, but then they argue that contemporary ecology is an ecology of *instability* that rejects equilibrium and balance.[7] So, like Regan, Hettinger and Throop reject Leopold's land ethic on the basis of two sentences alone.

---

6. The claim is originally due to Callicott (1987), and it has been repeated many times since by many authors, with the phrase "summary moral maxim" producing about 100 hits on Google Scholar as of January 2023. Indeed, as shown further below, a number of these myths have their origins in Callicott's work, even though he himself has subsequently sought to debunk at least one of them (namely, Myth 3). Callicott, who has published numerous essays and books on Leopold, has been called the "leading philosophical exponent of Aldo Leopold's land ethic" (Norton 2002, 127)—with no challenges to that ascription of which I am aware—and he has, for example, had an entire book devoted to discussing his views on Leopold (*Land, Value, and Community: Callicott and Environmental Philosophy*). But to be clear, the point of this section is not to criticize Callicott but to rectify widespread and persistent misunderstandings concerning Leopold.

7. Whether this understanding of contemporary ecology is fully correct—and I have my doubts—is separate from the point at hand.

Even scholars who are sympathetic to the land ethic seemingly endorse this myth. For example, Holling and Meffe (1996) use the two sentences as a jumping-off point to develop what they call a "Golden Rule" of resource management. They replace *stability* with *resilience*, but otherwise maintain that the two sentences constitute "sound advice."

Despite the ubiquity of the belief that these two sentences are a good summary of the land ethic, this belief is a myth that should be rejected. Leopold published over five hundred distinct items over the course of his lifetime; these are two sentences out of one essay in one book, published posthumously, and Leopold died before intended revisions to the book could be done (Meine 2010). We need to recall that Leopold was a scientist and not a philosopher, and that these words appear in an evocative essay written for a general audience; thus they should not be read literally. (In fact, this goes for all of ASCA). Instead, we need to consider the rest of "The Land Ethic," the context of Leopold's life experiences, and his statements elsewhere. When one does so, it becomes clear that Leopold expanded on these themes in a variety of ways and in a variety of contexts, sometimes using different words and in some cases changing his views as he reflected on his experiences. This casts a different light on the words in those two sentences. This contextual interpretative practice is standard in the history and philosophy of science, but it is less common in environmental ethics and conservation biology.

For example, read literally, it might appear from these two sentences that *anything* that benefited the integrity, stability, or beauty of a biotic community would be ethically right, even if it meant sacrificing the rights of individuals to do so. For this reason, Leopold has been accused of endorsing "environmental fascism," which would sacrifice individuals for the good of the community. However, other statements Leopold made do not support this interpretation; read literally, the two sentences do not accord with the rest of what Leopold says. Instead, the two sentences are meant to be emotionally resonant. The literal reading has given rise to the myth that Leopold is an environmental fascist—a myth that is debunked in the section headed "Myth 3" below.

Some readers see the words *stability* and *biotic community* as central to the purported summary moral maxim but fail to find explicit definitions of those terms within "The Land Ethic." They then seek to interpret Leopold in light of meanings used by ecologists of his time or ecologists of today: they assume that the terms are to be understood in some commonly used way. But such readers overlook the wealth of other places (which include other sections of "The Land Ethic" itself) where there are passages that are relevant for understanding Leopold's distinctive meanings. The following section, "Myth 2," discusses the meaning of *biotic community* and the section titled "Myth 5" discusses the meaning of *stability*. Finally, some readers might think, since the summary moral maxim doesn't mention humans explicitly, that we are not included. But again, this overlooks the extensive attention that Leopold gave to human practices and their role in biotic communities, which is discussed under the heading "Myth 4."

But these are more than four individual mistakes. The overall mistake is the assumption that the two sentences exhaust the land ethic without need for any further interpretive work. Once you reject this myth (Myth 1), then Myths 2, 3, 4, and 5 are quickly cast into doubt with just a bit of further examination. Myth 1 is in this sense a "keystone" myth.

Myth 2: When Leopold said "biotic community," he meant "ecosystem."

Leopold's purported "summary moral maxim" refers to the "biotic community," and it is widely believed that by "biotic community," Leopold meant "ecosystem." For example, J. Baird Callicott, while acknowledging the influence of Charles Elton's community concept on Leopold, suggests that it is ultimately "the physics-born ecosystem model" that Leopold turns to in "The Land Ethic" (Callicott 1989, 107). There are other authors who write as though Leopold was referring to ecosystems as the focus of the land ethic (e.g., Hettinger and Throop 1999; Knight 1996; and Vucetich et al. 2015), and it is less clear why these authors equate *biotic community* with *ecosystem*. Perhaps these authors are simply interpreting the purported "summary moral maxim" in what they take to be contemporary terms (similar to what seems to have happened with *stability*; see "Myth 5"). Yet, when one rejects Myth 1 and instead interprets the meaning of *biotic community* in light of what Leopold said elsewhere, a more complex picture emerges.

Leopold did state in "The Land Ethic" that land "is not merely soil; it is a fountain of energy flowing through a circuit of soils, plants, and animals" and that "[f]ood chains are the living channels which conduct energy upward; death and decay return it to the soil" (1949, 216). By including (what we would typically call today) abiotic components as well as matter and energy flow, there is indeed *some* reason to think that for Leopold, "biotic community" was just another way of saying "ecosystem."[8]

However, this myth should be rejected, because Leopold *also* stated that a biotic community is composed of interdependent species—that the biotic pyramid is composed of a complex tangle of lines of dependency for food and other services. And this makes his view sound similar to what today would be called an ecological or biotic community concept. Since interdependence plays a central role in the land ethic, there is no reason to think that Leopold "turned away" from the community concept, as Callicott suggests; a more plausible reading is that Leopold utilized a concept that incorporates aspects of *both* an ecosystem concept and a community concept. It is also worth noting that the term *community* nicely conveys the idea of an entity that we are a part of and connected to—and thus have moral obligations to—in a way that the term *system* does not. That is, the idea of community has the moral connotations that Leopold was seeking. So, Leopold does not "turn away" from the community concept; instead, he embraces and enhances it.

Leopold's term *biotic community* thus blended the concept of *ecosystem* from ecosystem ecology with the concept of *community* from community ecology (see chapter 3 for further discussion). That he did so makes sense when one considers that in the late 1930s and 1940s, both concepts were still fresh, emerging, evolving, and beginning to intersect with each other. Thankfully, Leopold used the term *land community* interchangeably with *biotic community*, so we can use the term *land community* to refer to the blended concept, reducing confusion—as I will do for the majority of this book. Moreover, there are contemporary analogues that combine ecosystem and community concepts that can be used to further elaborate Leopold's land community concept (Millstein 2018).

---

8. The term *ecosystem* was coined by Tansley (1935).

Here a concern might be raised about Leopold's use of the term *community*, given its origins in the thinking of Frederic Clements, an ecologist who is somewhat notoriously known for (among other things) characterizing communities as organisms (Clements 1916). Those raising such a concern might think that Clements's community concept has been superseded by the approach taken by ecologists Henry Gleason (1917) and Arthur Tansley (1935), challenging the claim made here that Leopold's land community is consistent with a contemporary understanding. However, Eliot (2011) has given good reason to think that Clements's commitment to communities as organisms has been overstated, and that Clements and Gleason, both having been interpreted in extreme terms, are actually not all that far apart in their views. In particular, Eliot demonstrates that, for Clements, communities aren't literally organisms; rather, they are *comparable* to organisms in certain (not very controversial) respects and not comparable in other respects. Moreover, by the end of his life, Leopold was de-emphasizing the community-as-organism view, and it plays a very small role in *A Sand County Almanac*. So, although community-as-organism might be an interesting idea for someone to pursue in thinking about environmental ethics and conservation biology, it is not a necessary aspect of the land community or something that contemporary community ecologists who claim to study communities subscribe to. The necessary component is only that there are interactions and interdependencies between components of the land community (see chapter 3), and even the most Gleasonian of ecologists acknowledges their existence.

It is important to reject this myth, because its rejection implies that ethical interactions with our land communities and conservation policies should seek to preserve not only matter and energy flow but also important ecological interdependencies and relationships, such as predator-prey, pollinator-pollinated, and so forth. Keystone species assume a particular importance.

Myth 3: Ecosystems are the only entities of value in the land ethic.

As noted in the discussion of Myth 1, some believe that according to the land ethic, biotic (land) communities are the *only* entities of value, giving rise to the understanding of the land ethic as a holistic ethic.[9] Again, taken at face value and out of context, the supposed "summary moral maxim" seems to define "what is right" entirely in terms of how we treat biotic (land) communities. Thus, it appears to endorse the sacrifice of individual organisms for the sake of the whole; for that reason, it has been called "fascist," as noted under "Myth 1."

However, this myth flies in the face of many other statements that Leopold made. For example, he clarified that the "land ethic... implies respect *for his fellow-members*, and also respect for the community as such" (Leopold 1949, 204; emphasis added). This is an explicit denial of the claim that only the biotic community matters; rather, individuals and the community *both* deserve our respect. Similarly, he maintained that individuals (wildflowers, songbirds, predators) need not have an economic value or even a functional value in the land community in order to continue as a matter of "biotic right"—that "no special interest has the right to exterminate them for the sake of a benefit, real or fancied, to itself" (Leopold 1949, 211).

Furthermore, in "The Land Ethic," Leopold states that he saw the history of ethics as a history of "accretions," beginning with relations to individuals, then expanding to include the relation between individuals and society; the land ethic, Leopold suggests, would be a third accretion. He also refers to the land ethic as an "extension of ethics" (1949, 128). Leopold's wording here implies that our ethical obligations to the land would not supersede our obligations to individuals, but would add to them. Again, this challenges the accuracy of Myth 3.

Rejecting this myth prevents the overly quick rejection of the land ethic as fascist.[10] However, it does make our conservation policies harder to craft, because we will have to balance the rights of

---

9. Callicott (1987, 196) states that not only does the land ethic have "a holistic aspect" but that "it is holistic with a vengeance." In a subsequent work, Callicott recants this view, stating that Leopold never meant the land ethic to completely override all of our duties to other humans (Callicott 1999). However, the earlier paper may have had some lingering influence, despite Callicott's recanting.

10. For philosophical defenses against the "fascism" charge, see, for example, Nelson 1996, Marietta 1999, and Callicott 1999. Meine's (2010) thorough discussion of Leopold's life and work, political beliefs and activities, and familial and ethical background, makes clear that there is no historical substance to the fascism charge.

individuals against the rights of the land community as a whole. If Leopold is right, that is a balance worth striving for, even if sometimes impossible to fully achieve in practice. These issues are discussed further in chapter 6.

Myth 4: *The central message of the land ethic is to set aside human-free ecosystems.*

Some seem to believe that the central message of the land ethic is to set aside human-free ecosystems. For example, Laura Westra sees the land ethic as applying to "largely undiminished and unmanipulated natural systems" (Westra 2001, 263). Rohlf and Honnold state that "[t]o Leopold, wilderness was the land ethic's ultimate expression—an interdependent biotic community unimpaired by human manipulation" (Rohlf and Honnold 1988, 254). Guha (1989) seems to have understood Leopold similarly.

However, this myth should also be rejected. Leopold was explicit in including humans as parts of many food chains in many land communities; he emphasized human interdependence with other species and with the abiotic components of the land community (see chapter 2 for further discussion). Indeed, Leopold spent much of his career not only trying to institute sound forestry, wildlife management, and farming practices, but also actually working to integrate these practices (Meine 2010). This is clear even in "The Land Ethic" itself, where, for example, Leopold discusses the need for farmers to value the land—including privately owned land—and to feel an obligation toward the land in order to institute and *maintain* practices that preserve the soil. Forestry is also discussed explicitly. Thus, the land ethic encompasses all of these human practices, emphasizing how we should live on the land and not merely trying to set it aside.

In rejecting this myth, it becomes clear that the main point of the land ethic is *not* to set aside reserves where no humans tread, although Leopold did argue that there are reasons to do that in certain regions. He recognized that "[m]any of the diverse wildernesses out of which we have hammered America are already gone" (1949, 121), but he thought that there were remnants of varying sizes and degrees of wildness, and that a "representative series of these areas can and should be kept" (1949, 122). He gave several reasons for preserving wilderness: (1) for recreation, in order to perpetuate "in sport form, the more virile and primitive skills in pioneering travel and subsistence" (1949, 123); (2) for science, in order to have a "base datum of normality, a picture of how healthy land maintains itself as an organism" (1949, 125), so that our conservation and restoration efforts have a greater chance of success and so that we know what "success" looks like; and (3) for wildlife, which require large areas, larger than the national parks in the United States. So, Leopold clearly did think that some wilderness should be set aside. The point of this subsection—why this myth should be rejected—is rather that setting aside wilderness is not the central or sole focus of the land ethic.[11]

When we reject the myth that setting aside human-free ecosystems is the central focus of the land ethic, it becomes clear that *all* of our human practices matter—that we always need to think about our effects on other species and their effects on us. Modifying our human practices can be important conservation efforts, too, whether via the actions of individuals or enshrined in policies.

Myth 5: By "stability," Leopold meant something like "balance" or "dynamic equilibrium."

The fifth myth, which was discussed briefly in the context of Myth 1, is that by "stability," Leopold meant something like "balance" or "equilibrium." As Eric Freyfogle (2008) points out, many commentators quote the purported "summary moral maxim," but few try to figure out what *Leopold* meant by "integrity," "stability," and "beauty." With respect to stability in particular, Freyfogle suggests that these authors simply assume that Leopold meant that land communities should be static or unchanging,[12] or, like Callicott, they try to assimilate Leopold's meaning to that of other ecologists.

However, Leopold did not use the term *stability* the way other ecologists of his time did. Leopold explicitly studied changing

---

11. See Meine (2010) for further discussion that traces Leopold's changing views on the issues discussed in this section.

12. This might seem an unlikely view for any biologist to hold, and indeed, dynamic equilibrium is probably a more common view, but it is not unheard of. See, e.g., Whittaker (1999).

ecosystems, such as the effects of fire and drought (Meine 2010). And he often contrasted slow, mild changes that land communities could adjust to, with rapid and drastic changes that led to dust-bowl type situations; characterizations of these contrasts can be found within the "Land Ethic" essay itself.

Instead, as Julianne Warren (2016) persuasively demonstrates, Leopold's idea of stability meant something closer to "land health": the capacity of the land for self-renewal, so that it continues to sustain life over time. And this moves Leopold's understanding of stability a lot closer to contemporary terms like *sustainability* or perhaps *resilience* (see chapter 4 for further discussion).

This improved interpretation avoids the hasty rejection of the land ethic for purportedly using an outdated notion of stability. It also directs us to consider actions that preserve or enhance self-renewal and thus land health, such as preserving soil health, preventing the extinction of species (preserving "integrity"), performing appropriate restorations, and making any changes *carefully*.

Myth 6: Leopold's ethics are derived from Charles Darwin's "proto-sociobiological" perspective on ethical phenomena.

The sixth and last myth is the claim that Leopold's ethics are derived from Charles Darwin's "proto-sociobiological" perspective on ethical phenomena. According to an influential interpretation by Callicott (1987; 2014), Leopold, drawing on Darwin's account of ethics in the *Descent of Man*, believed that humans evolved to have bonds of "affection and sympathy" toward humans who were not relatives because those bonds conferred advantages on communities that contained such individuals. As such humans became ecologically literate, these "moral sentiments" would be "automatically triggered" toward the biotic community, thus conferring moral value on biotic communities (Callicott 1987, 194). (This is a broadly Humean, and somewhat controversial, approach to ethics).

This myth, like the others, should be rejected.[13] Callicott's pri-

---

13. See Millstein (2015) for a fuller discussion of the arguments in this subsection. Also, note that Norton (e.g., 1988; 2005) argues for a competing interpretation of the ethical basis for the land ethic; he maintains that Leopold was influenced by American Pragmatism rather than Darwinian/Humean ethics. I am not taking a

mary evidence that Leopold is drawing on Darwin's account of the evolution of ethics in the *Descent of Man* is Leopold's use of the phrase "struggle for existence." However, "struggle for existence" is an idea developed in the *Origin of Species*, not the *Descent of Man*; it is the title of chapter 3 of the *Origin*, where Darwin discussed the interdependencies among organisms in the struggle for existence (Millstein 2015). The struggle for existence is more commonly associated with the competition between organisms for survival, but in the *Origin*, Darwin clarified that this struggle for life is broader than competition, including, for example, a struggle to survive in the face of difficult climatic conditions. Darwin further pointed out that organisms (usually the more distantly related ones) that are engaged in a struggle for existence in fact *depend* on each other for survival, as do a bumblebee and a flower. Interdependence in this sense is a core theme of "The Land Ethic," and many of Leopold's phrases echo Darwin's from the *Origin* (Millstein 2015).

The rejection of this myth reveals that the land ethic is not dependent on the vagaries of human sentiment. Rather, the basis for the land ethic derives from our interdependencies with other organisms, suggesting (again) that the focus of our conservation efforts should be on understanding, preserving, and (when relevant) restoring the interactions between organisms in a land community in order to maintain, promote, or restore land health.

Of course, Leopold thought that our feelings toward other organisms and toward the land community were relevant to how we would in fact behave toward it; he makes this point a number of times and in a number of places. My claim is only that there is no evidence that he thought those moral sentiments formed the *basis* for our obligations—that is, there is no evidence that he thought we would have no ethical obligations without those sentiments. Rather, the textual evidence suggests that Leopold thought our interdependence with other members of the land community is the

---

stand on Norton's claim because I think it turns on what one means by American Pragmatism, a question that would take the discussion far afield from the goals of this book. I certainly agree with Norton that Leopold was in some sense a "pragmatist" (whether or not he should be considered an American Pragmatist, or heavily influenced by American Pragmatism), and I refer to him as such at various points in this book. My alternative interpretation of his ethical basis appears in chapter 5.

basis of our obligations toward it, regardless of our feelings (but, again, our feelings do serve to motivate us to act).

Reject the Six Myths

Accepting all six myths described above entails accepting a distorted picture of Leopold's beliefs—one where individuals are sacrificed to the good of the ecosystem, characterized in terms of its matter and energy flows and where the "good" of an ecosystem is understood in terms of outdated and unrealistic concepts of stability. It also means accepting a view where the only goal is to set aside ecosystems completely free of human encroachment, all of which is predicated on humans extending their moral sentiments (fellow feelings) to ecosystems.

Rejecting all six myths and accepting the alternative interpretations presented in this chapter entails accepting a picture of Leopold in which he values both individuals and the land communities they are a part of, with land communities consisting of interacting interdependent organisms, abiotic components, and matter/energy flow; in which he understands the "good" of a land community in terms of its health, characterized in terms of its ability to continue the nutrient cycling necessary to sustain life over time; in which our numerous goals include maintaining important ecological relationships and matter/energy flows, preserving soil health, and preventing the extinction of species, all of which is predicated on the fact that humans and other species are interdependent with each other, so that their fates are not separable. This is an appealing, practical, and moderate picture of the land ethic that can guide our actions, and it yields a more defensible and fruitful ethical basis for conservation policy.

In the rest of the book, I do not treat the so-called summary moral maxim as any more central than anything else that Leopold has written concerning the land ethic. It should be understood as more evocative than literal and somewhat misleading when read literally.

OVERVIEW OF THE BOOK

The book is structured as follows. Chapters 2, 3, and 4 each take up a central concept in Leopold's land ethic: interdependence, land community, and land health, respectively. These chapters might appear to fall within the history and philosophy of biology, although all three concepts have been analyzed within environmental ethics and ecology/conservation biology as well, and thus the chapters speak to and engage with those areas too. Chapter 5 draws on those three clarified and defended concepts in order to spell out Leopold's argument (actually, arguments) in favor of the land ethic. It is perhaps the chapter that falls the most squarely within environmental ethics. With the land ethic defended, chapter 6 shows the role that it can play in policy.

Each chapter begins with a central case that is referred to throughout the rest of the chapter. These cases, and other cases discussed more briefly within each chapter, are meant to help illustrate the issues at hand as well as to show the real-world applicability of the land ethic. Thus, some empirical challenges arise along with the conceptual ones: for example, are there entities in nature cohesive enough to deserve the name *land community*? Are the interactions between species populations strong enough to make interdependence ethically significant? Do the cited studies from which the cases are drawn actually show what they purport to show? Given the nature of this book, which is meant to be a work in the history and philosophy of ecology and environmental ethics (albeit informed by the science of ecology), I cannot tackle these empirical challenges head on—that is, I cannot (and do not claim to have the expertise to) provide empirical evidence that would sway the reader. These are issues about which ecologists themselves disagree. Of course, much of science is controversial, so I need to plead for a sympathetic reading on any controversial empirical issues, with the understanding that I have to make some assumptions in order to make any progress at all, and with the understanding that all scientific claims are subject to being overturned with new and better evidence—something that Leopold himself was keenly aware of and frequently remarked on. So I put forward my arguments with the assumption that we need to proceed humbly and cautiously.

Some chapters contain more historical interpretation and others more philosophical analysis, depending on the extent to which Leopold discussed each topic and provided his own defense. I draw on Leopold's published and unpublished works, emphasizing those from later in his life but not excluding his earlier work. This is because Leopold modified and elaborated many of the ideas discussed in this book over the course of his lifetime. In some cases, his earlier works are informative for understanding the later ones, but in other cases, he has changed his views (e.g., regarding the hunting of predators).

I now turn to brief summaries of each of the chapters that follow.

Leopold stated that "All ethics so far evolved rest upon a single premise: that the individual is a member of a community of interdependent parts" (1949, 203). *Interdependence* forms the backbone of the land ethic, so its proper understanding is essential. Chapter 2 thus focuses on characterizing a Leopoldian conception of interdependence. Interdependence might be thought to consist only of positive interactions, but I argue that, for Leopold, both positive and negative interactions between species populations (such as competitive interactions or the interactions between parasite and host) give rise to interdependence. Importantly, human practices are explicitly included among the relevant interactions (since humans are parts of land communities). Moreover, abiotic components of a land community are interdependent with other parts of the community as well. I provide a defense of Leopold's conception of interdependence, using both philosophical analysis and examples from contemporary ecology. I further show how interdependence can be understood at multiple scales, including a global scale. Finally, I suggest that interdependence is best understood in the context of a web (or network) of interactions, which leads to the topic of *land community*.

As noted above, some authors have interpreted Leopold's "land community" to be synonymous with Clements's meaning of community, and on that basis dismiss his view as an outdated ecological idea from the early twentieth century. Other authors assume that Leopold meant something closer to ecosystem. In chapter 3, I give evidence that Leopold's conception of a land community was a blending of these two ideas, a conception that is perhaps not embraced by the majority in contemporary ecology but which has strong contemporary support. It is also sometimes argued that land communities need to have delineable boundaries in order to be entities of value, as the land ethic posits, and I show that we can characterize their boundaries in terms of relative strengths of species population interactions and matter/energy flow. But being an entity with delineable boundaries is not sufficient for land communities to be entities of value; it is also commonly thought that entities must be capable of being benefited or harmed in order to be entities of value. That raises the subject of *land health*.

In chapter 4, I elaborate and defend Leopold's conception of land health—a desirable state, according to the land ethic. In "The Land Ethic," Leopold described land health as "the capacity for self-renewal." It was an idea he had been working on for some time that was still in process at the time of his death, so understanding it requires a bit of sleuthing. I begin with an exploration of what Leopold meant by land health and what he saw as its symptoms, causes, and underlying mechanism. Underlying the land health capacity, Leopold believed, was the complex web of food chains (made up of species interactions and resulting interdependencies) arranged in a metaphorical land pyramid, with energy and nutrients flowing up the pyramid from the soil to the top predators and eventually returning to the soil. Leopold hypothesized that with longer food chains, land communities could persist—sustain biodiverse land communities for longer periods of time—because nutrients would be bound up in organisms rather than in the soil, where they were always at risk of loss through wind or water erosion. After characterizing Leopold's conception of land health, I compare Leopold's approach to contemporary debates over the connection between biodiversity and stability, finding points of similarity with some research and dissimilarity with others. I then address potential concerns regarding Leopold's conception of land health and discuss its philosophical and scientific significance.

One question still to be addressed, though, is why we humans ought to act to protect or promote the health of land communities. In chapter 5, I argue that Leopold's main argument for the land ethic rests upon seeing that most people already accept the basic principles on which a land ethic can be justified—accepting obligations and rules that limit our freedom of action because we are interdependent with other humans. Once we examine what justifies and grounds our human ethics, and once we understand land communities properly and our interdependence with other members of land communities, we should see that the same principles justify and ground extending obligations to the land. The chapter elaborates this argument further (and other related arguments), and then identifies various explicit and implicit premises underlying the land ethic, showing that each is at least plausible and reasonable. These premises have to do with the extent of interdependence assumed by the land ethic, whether land communities are entities that should be included in our moral deliberations (their "moral considerability"), the ethical basis for the land ethic, human-only and human-plus-nonhuman interdependence, what a complete account of our moral obligations includes, and the intrinsic value of land communities.

If the arguments of the previous chapters have been successful, then the land ethic is a defensible basis for our actions. But can it be more than that—can it be a guide for our environmental policies? I answer this question in chapter 6, using Leopold's own policy-related activities and his stated reflections on them to elaborate on what he called "the Principle of Integration of Land Uses." I show how this approach sought to bring together many different interests, values, and perspectives (and their corresponding methods) to cooperate rather than compete—to find harmonious and balanced solutions to particular issues while accepting the common purpose that makes all other interests possible: namely, the health of the land. To further spell out this approach to policy—which spells out a policy process rather than dictating certain types of outcomes—I characterize eight subprinciples of the Principle of Integration of Land Uses. I then clarify and elaborate what I take to be the two biggest challenges of implementing these subprinciples: the challenge of balancing values and interests and the challenge of incorporating land health into policy. Finally, I address other potential issues of concern, such as the question of the extent to which humans should take an active role in maintaining "natural areas," the extent to which Leopold was committed to setting aside "wilderness," and whether the Leopold-inspired policy process I have outlined is substantive enough to provide guidance.

In sum, this book seeks to show that Leopold's land ethic is a reasonable, plausible, and practical approach to dealing with the nonhuman parts of the world, one that is consonant with prominent approaches in contemporary ecology. At the very least, I hope to convince you that the land ethic is a candidate for environmental ethics and environmental policy that is worthy of our consideration. If I am fully successful, I will have shown you that the land ethic is the environmental ethic for our time.

CHAPTER TWO

Interdependence

All ethics so far evolved rest upon a single premise: that the individual is a member of a community of interdependent parts. . . . The land ethic simply enlarges the boundaries of the community to include soils, waters, plants, and animals, or collectively: the land.

ALDO LEOPOLD, *A Sand County Almanac*

Parts of this chapter originally appeared in Roberta L. Millstein, "Understanding Leopold's Concept of 'Interdependence' for Environmental Ethics and Conservation Biology," *Philosophy of Science* 85 (2018): 1127–39. © 2018 by the Philosophy of Science Association. All rights reserved.

INTRODUCTION: WOLF-DEER-PLANT INTERDEPENDENCE

In the early part of his life, Aldo Leopold, a hunter of deer and other animals, thought that the way to have more deer available was to kill off the animals who were preying on them. To this end, Leopold shot wolves and other deer predators. Indeed, in the early twentieth century, the US government had an official policy that sought the eradication of wolves. (It came close, but it did not "succeed"). But later in his life, faced with populations of starving deer and denuded trees, Leopold was forced to realize the error of this approach. In one of his most well-known essays, "Thinking like a Mountain," he wrote:

> I have lived to see state after state extirpate its wolves. I have watched the face of many a newly wolfless mountain, and seen the south-facing slopes wrinkle with a maze of new deer trails. I have seen every edible bush and seedling browsed, first to anaemic desuetude, and then to death. I have seen every edible tree defoliated to the height of a saddlehorn. Such a mountain looks as if someone had given God a new pruning shears, and forbidden Him all other exercise. In the end the starved bones of the hoped-for deer herd, dead of its own too-much, bleach with the bones of the dead sage, or molder under the high-lined junipers. (Leopold 1949, 130–32)

His study of the wolves and deer on the Kaibab Plateau in Arizona exemplified this new attitude. In "Deer Irruptions," Leopold described a formerly stable Kaibab deer herd of around 4000 deer that began to increase around 1910, with the range showing overbrowsing. According to Leopold, by 1924 the deer herd had increased to 100,000, followed by a famine that reduced the herd by 60 percent over two winters. By 1939, Leopold estimated that the herd was down to about 10,000, with a lowered carrying capacity. Leopold cited the loss of predators (cougars and wolves) along with fire control as events that paved the way for deer irruptions, and he compared the Kaibab to similar locations across the United States with similar dynamics.

For a time, based on Leopold's analysis, the case of the Kaibab was cited in numerous ecology textbooks as an exemplar of the dangers of predator removal and of a *trophic cascade* (Young 2002). A trophic cascade is a situation in which "the presence of top trophic-level predators significantly affects herbivores (the next lower trophic level), and this interaction alters or influences vegetation (e.g., species composition, age structure, or spatial distribution)" (Ripple and Beschta 2005). Leopold's account was challenged by Graeme Caughley in 1970, causing the example to be removed from many textbooks. Then, thirty-six years later, Dan Binkley et al. (2006) successfully debunked Caughley's debunking and vindicated Leopold's original conclusions with a thorough re-analysis of the evidence (Millstein 2019).

The case of the Kaibab is not only a canonical example of a trophic cascade; it is also a canonical example of *interdependence*. (Indeed, all trophic cascades require interdependence). Wolves obviously depend on deer for food, and deer obviously depend on aspen or other vegetation for food. What the case of the Kaibab shows is that deer are also dependent on wolves. Without wolves or other predators to keep the size of deer populations in check, deer will eat all the available food and starve to death.

And it's not just "charismatic megafauna" like wolves and deer that exhibit interdependence. Here are some other cases where populations of different species seem to depend on each other: humans and the bacteria in our guts, with bacteria not only doing things like producing vitamins and digesting our food, but also potentially influencing our behavior and cognition (Lyte 2014); interactions between wild bees and flowering plants via pollination, also seemingly affected by landscape composition and habitat richness (Papanikolaou et al. 2017); the effects of trees on human health related to cardiovascular and lower-respiratory-tract illness (Donovan et al. 2013) and the dependence of trees, at least in urban areas, on humans for care and maintenance; and a complicated trophic cascade including benthic fish species, grey seals, small pelagic fish species, benthic macroinvertebrates, zooplankton, and phytoplankton (Frank et al. 2005). Other examples of interdependence are discussed below.

As the epigraph to this chapter suggests, interdependence formed the central basis for Leopold's land ethic. Thus, understanding interdependence and determining its conceptual coherence are essential for understanding the land ethic and determining its conceptual coherence. It should be noted that Leopold was neither the originator of the concept of interdependence nor its sole explicator; his views on interdependence were influenced by Charles Darwin and Charles Elton, and he influenced prominent ecologists like Eugene Odum in turn. (See chapter 1 and Millstein 2015 for more on Darwin's influence in particular.).

Subsequent chapters elaborate on why interdependence matters and the role that it plays in the land ethic and its justification. But the rough idea is this: just as our interdependence with other humans has led us to recognize the value of human communities and our obligations to those communities, our interdependence with other species and abiotic features should lead us to recognize the value of land communities. Interdependence is why we should care, why we have ethical obligations beyond humans and individual organisms.

I begin this chapter with an overview of Leopold's conception of interdependence. This prompts a series of questions: Are humans included in communities of interdependent parts? Does interdependence consist only of "positive" causal interactions, or are "negative" interactions included as well? What makes an interaction "positive" or "negative"? What entities are involved in causal interactions, and what do the interactions affect? The answers to these questions lead to some interim suggestions for how we should understand interdependence. I then respond to two possible objections to those interim suggestions by further elaborating and defending the concept of interdependence. My overall goal is to produce a defensible conception of interdependence that is as Leopoldian as I can make it (that is, that draws on his known views on interdependence, seeking to make explicit what is sometimes implicit) while keeping it conceptually coherent and consistent with contemporary ecological findings. To that end I conclude with a proposal for the concept of interdependence for the land ethic that, given Leopold's influence, should also be relevant for related areas such as conservation biology, forestry, wildlife management, and restoration ecology.

WHAT DID LEOPOLD MEAN BY "INTERDEPENDENCE"?

By the end of his life, Leopold was characterizing interdependence in terms of food chains, or "lines of dependency for food and other services" (Leopold 1949, 215); see figure 2.1. His notion of a food chain—the sequence of stages in the transmission of food, established by evolution (Leopold 1942a)—is tied to his conception of a land pyramid, where "each successive layer depends on those below it for food and often for other services, and each in turn furnishes food and services to those above" (1949, 215); see figure 2.2. Both the food chain metaphor and the land pyramid metaphor show Charles Elton's (1927) influence on Leopold (see Elton 1927 and Warren 2013 for discussion). An example food chain is a squirrel that drops an acorn, which feeds a quail, which feeds a horned owl, which feeds a parasite (Leopold 1942a, 205). But there are other "chains of dependency" in addition to those involving food: "The oak grows not only acorns; it grows fuel, browse, hollow dens, leaves, and shade on which many species depend for food and cover or other services" (205). The land pyramid, Leopold states, contains a "tangle" or "maze" of all these types of chains.

Moreover, it is not just biotic components that form these interdependencies; as figure 2.2 shows, soil is at the base of Leopold's land pyramid, and he saw plants such as oak trees as dependent on soil, with all food chains ultimately returning some of their matter and energy back to soil. I return to the topic of the interdependence of abiotic components below. In what follows, I refer to causal relationships, such as feeding relationships, between organisms of different species or between different species populations,

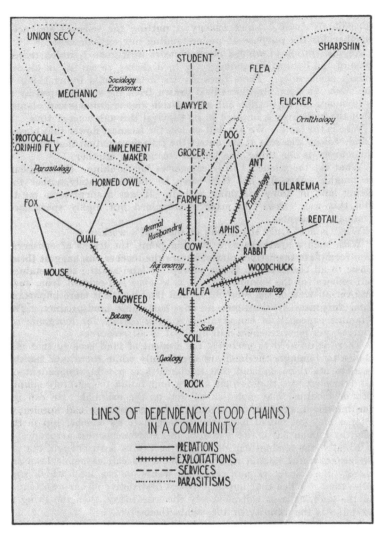

FIGURE 2.1. Leopold's 1942 depiction of the tangled "lines of dependency"—food chains—in a land community. What Leopold called food chains comprise diverse types of interactions, not all of which are trophic, and require study from diverse areas of inquiry (e.g., sociology, botany, animal husbandry, geology, etc.). From Aldo Leopold, "The Role of Wildlife in a Liberal Education," *Transactions of the Seventh North American Wildlife Conference*, 8–10 April 1942, 485–89. Copyright 1942 by the Wildlife Management Institute. Reprinted with permission.

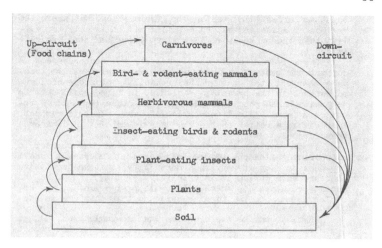

FIGURE 2.2. Leopold's 1939 depiction of a biotic/land pyramid. Leopold envisioned that energy would be conducted up the pyramid from the soil and eventually return toward the soil via food chains. From Aldo Leopold, "A Biotic View of Land," *Journal of Forestry* 37 (9): 727–30. Courtesy of the Aldo Leopold Foundation and University of Wisconsin–Madison Archives, https://search.library.wisc.edu/digital/AG65AV6OBR2TSI8G/pages/AOFUK2MXMTUS4382. Reprinted with permission.

as *interactions*.[1] The suggestion is that these interactions give rise to (yield, generate) interdependencies.

Are Humans Interdependent with Other Members of the Land Community?

Leopold states explicitly that humans and their agricultural products are parts of these food chains and are thus interdependent with other biotic and abiotic elements: "Each species, including ourselves, is a link in many chains. The deer eats a hundred plants other than oak, and the cow a hundred plants other than corn. Both, then, are links in a hundred chains" (1949, 215).

---

1. Leopold also uses the term *interactions*, stating that they occur between "components of the land" (i.e., "soils, water systems, and wild and tame plants and animals"; 1942a, 199).

However, some environmental ethicists have challenged this claim of human interdependence: "We are undoubtedly dependent on them, but in what ways are ecosystems dependent on us? Their independence from us is not like the independence of parents from offspring who can later reciprocate love and other mutual activities that can develop into interdependency. We play no such role in any ecosystem; we seem genuinely superfluous to ecosystem functioning" (Ouderkirk 2002, 6; see also Taylor 1981). Ouderkirk seems to be implying that interdependence is only about positive interactions—reciprocal, mutually beneficial interactions—and that humans do not engage in these with other species.

But humans *sometimes* have positive effects on other organisms. For example, humans have positive effects on corn, sheep, squirrels, rats, and pigeons—or at least on their population sizes. And some of these organisms have positive effects on us—again, at least on our population sizes. Corn, for example, serves as a staple in many human diets.

Does Interdependence Result Only from Positive Interactions?

Even if one grants that humans can have positive effects on other species, a further question needs settling: namely, does interdependence really include only positive, mutually beneficial interactions? Ecologists typically identify a variety of causal interactions by their positive and negative effects. Consider the following examples:

competition—negative for both sides
amensalism—negative on one side, neutral on the other
parasitism, predation—positive on one side, negative on the other
commensalism—positive on one side, neutral on the other
mutualism—positive for both sides

This list of interactions is not exhaustive. For example, it leaves out a large class of positive-for-human-negative-for-nonhuman interactions, ones where humans are not so much acting as parasites or predators but engaging in more wholesale destruction, as when we destroy habitat for commercial development or drive other species to extinction. These interactions might be better characterized as *exploitation*, a term that Leopold used in the diagram shown in figure 2.1 and elsewhere in his writings. For example, he referred to "exploitative agriculture" as agriculture that has exceeded sustainable carrying capacity (Leopold 1949). Since exploitations present a particular challenge for interdependence, I discuss them separately below (see the subsection below headed "Objection 2").

Ouderkirk's suggestion seems to be that, of these types of causal interactions, only the mutualistic ones exhibit interdependence. Humans might have parasitic, competitive, or even commensalistic interactions with other organisms, but in Ouderkirk's picture, these would not amount to interdependence. Relatedly, Christopher Eliot (2011) seems to imply that competitive interactions in particular are not dependence relations.

However, Leopold explicitly considered *all* these causal interactions, both positive and negative, to exhibit interdependence. In the key for the figure reprinted here as figure 2.1, Leopold indicated that the "lines of dependency" can be predations, exploitations, services, or parasitisms. Again, one of the examples of interdependency that Leopold returned to frequently was the predator-prey relationship between wolves and deer, most famously in "Thinking like a Mountain" (Leopold 1949). Furthermore, in discussing the lines of dependency in "The Land Ethic," Leopold stated that the land pyramid's "functioning depends on the co-operation and *competition* of its diverse parts" (1949, 215; emphasis added).

Defenders of Ouderkirk's view would probably think that Leopold should not have included both positive and negative causal interactions in his concept of interdependence—that his conception of interdependence was muddled or confused. However, Leopold was right to consider that both negative and positive causal interactions can give rise to interdependence, for two reasons.

First, interactions cannot always be definitively characterized as positive or negative. Organisms can simultaneously exhibit negative and positive interactions, as when vascular plants compete for limited resources at the same time that they provide each other structural support (Harley and Bertness 1996). Or, they can be positive in one context but negative in another: for example, yeast strains change from being mutualistic to being competitive depending upon the amount of freely available amino acid in the environment (Hoek et al. 2016)—an abiotic component.

Second, causal interactions might be negative in one sense but positive in a different sense, especially when one varies the time scale. Consider again Leopold's example of wolves and deer on the Kaibab plateau. Initially, one might reasonably say that wolves had a negative effect on deer by preying on them and thus reducing their population size. But when wolves were eliminated from the Kaibab—deliberately extirpated by a government-sponsored killing program—the deer populations exploded. So, one might think that removing the wolves had a positive effect on the deer. Ultimately, however, without wolves to keep the deer population in check, the deer ate much of the available browse,[2] and many of them starved to death. Their population sizes crashed. Thus, wolves arguably had a positive effect on deer; in the presence of wolves, the deer were healthier and able to maintain a more stable, yet smaller, population size. These considerations suggest that not only was Leopold right to include negative interactions in his concept of interdependence, but also that the variability and context dependence of these interactions shows that rigidly classifying them as negative and positive can be problematic.

Returning to the case of humans, we can see that the claim that humans are interdependent with other species does not necessarily imply that human actions that have led to the increase of various species populations are unequivocally good; those populations often threaten other members of the land community and the health of the community as a whole. But those species are, for better or for worse, dependent on us, and our future actions will affect their future. Perhaps most importantly, Leopold wanted to highlight the ways in which human actions could benefit the soil (e.g., by using crop rotation) or harm the soil (e.g., by allowing overgrazing), and the often serious downstream consequences of those actions for other species.

Recent scientific findings corroborate Leopold's view that humans are interdependent with other organisms. For example, Thomas et al. (2009) show that the globally threatened *Maculinea arion* (Large Blue Butterfly) is adapted to a single host-ant species (*Myrmica sabuleti*) that was affected by human-initiated grazing practices. Changes in grazing practices caused a local decline in the ant

---

2. Leopold characterized "browse" as "twigs, buds, and catkins of woody plants" (1943, 357).

species, and thus a decline in the butterfly species too, since the latter was dependent on the former. Restoring the previous grazing practices reversed the decline of the butterfly. In other words, both the ant and the butterfly were dependent on human activities. More generally, Sullivan et al. (2017) argue that other organisms have been genetically adapted, via natural selection, to the presence of humans. Indeed, given that our actions have affected nearly every other species on this planet, it seems fair to say that most other species are dependent on us. This dependency does not mean that all of our potential actions will improve the situations of other species any more than deer browsing improves the situations of tree seedlings, shrubs and forbs.

Still, one might reasonably ask whether certain species (human or otherwise) are more crucial than others, and thus question whether all dependence relations are equally strong. Species that are particularly important for the integrity and stability of the rest of the community are known as "keystone" species (Paine 1969). They are often predators, but not always; for example, pollinators are also seen as keystone species.

However, biologists have shown that whether a particular species acts as a keystone species depends crucially on the environment (e.g., Lubchenco and Menge, 1978). For example, one study showed that, in an intertidal zone, predators play a large role in protected areas but a small role in areas with large waves. Given that global warming is leading to climate change, and given that species interactions seem to be an important proximate cause of extinction in response to climate change (Cahill et al., 2013), it would be hasty to conclude that any particular species everywhere and always plays a greater (or a lesser) role than others. It is also worth pointing out that Leopold himself urged us to recognize our limited understanding of ecological systems and to act with caution. Interdependence is not eliminated or lessened by our identification of some species as keystone species in a particular environment at a particular time.

What Constitutes a Negative or Positive Causal Interaction?

The wolf-deer-plant example on the Kaibab plateau, by contrasting the health of individual deer with the control of deer population size, raises two further questions: First, what do ecologists mean when they say that some interactions are negative and some positive? Second, are the interactions negative for organisms, populations, or the whole land community?

The answers to these questions often go unspecified—and those who do specify often disagree. For example, Odum (1971) asserts that positive interactions result in population growth, whereas negative interactions produce a population decrease. Brooker and Callaghan (1998), on the other hand, characterize positive interactions as the increased "performance" of organisms, such as increased size, whereas negative interactions result in the decreased "performance" of organisms. So, in both cases, whether a causal interaction is characterized as positive or negative has to do with the *effect* or *outcome* of the causal interaction (and is thus not really about the *type* of causal interaction). But Odum considers the effects on *populations*, whereas Brooker and Callaghan consider the effects on *organisms*. And, of course, the relevant sorts of effects are different as well: increased numbers in the former cases, and increased "performance" in the latter.

However, the case of the Kaibab shows why sometimes "performance" is negatively correlated with population growth: the individual deer are healthier when their population sizes are smaller. So, the effect of the wolves on the deer (the result of a predator-prey relation) cannot be unequivocally characterized as "positive" or "negative." There were both positive and negative effects on individual organisms (positive for the wolves and healthier deer, negative for the deer who were killed) and negative effects for the population.

One possible response to the discrepancy between the meanings of positive and negative interactions would be to use them only in referring to effects on population size. That stipulation would be consistent with the focus of population ecology.[3] But the case of the Kaibab also shows that whether wolves positively or negatively affect deer population sizes depends on the time scale and other populations present (in this case, whether there was browse for deer to eat), since ultimately the deer population crashed without wolves. So, limiting our understanding of positive and negative to

---

3. See Molles 2015 for an alternative approach, where positive interactions are those that increase the fitness of individual organisms.

populations would only go so far in removing ambiguity. Recall that predation is typically characterized as positive for the predator and negative for the prey; the case of the Kaibab shows why this characterization is misleading.

Furthermore, there may be reason to interpret interdependence more broadly than just as the causal interactions between populations of different species. Leopold's concept of interdependence included effects on both the biotic and abiotic components of land communities. In a lecture in 1941, Leopold gave the following extended example of interdependence, in which abiotic components play a key role. A Wisconsin farmer wants more cows. To have more cows, he needs more corn and pasture, so he clears a slope—but he clears it too high. As a result, formerly small watercourses are now cut by gullies. These carry soil (and thus fertility) away and there is also flooding. The floods result in a loss of lowland pasture, the suffocation of trout, and the destruction of highways and railroads. Leopold asks, "who suffers?" His answer is that the farmer suffers; the farmer loses soil fertility, runs out of firewood, and is forced to buy coal. The neighbors below the farmer also suffer because they lose land and possibly buildings. Taxpayers suffer because they must pay for the flood damage in higher taxes and prices. Fishers suffer because there are no trout to catch; their choice is to fish for carp or stay home. But it's not just humans who suffer. Wildflowers and partridge are extirpated from the area because they can only live in unpastured woods. Woodcocks are similarly evicted because they inhabit only timbered streams—and so on. Ultimately, the chain of events leads to more rural slums and abandoned farms.

Leopold concluded, that "[t]his chain of evils, arising from one abuse affects *all* resources. The penalties of abuse are both economic and esthetic. They hit *all* people. Hence, I speak of the unity of land, and say that all parts of the land, including ourselves, prosper or decline together" (Leopold 1941a, 950).

Leopold's example shows what we'd be missing if interdependence were only to include the effects on the sizes of populations.[4]

---

4. While it is helpful in showing a chain of interdependencies that includes abiotic components, Leopold's example is limited in other respects. For one, it consists of uniformly negative effects. Other networks of interactions might contain a mix of positive and negative effects, as depicted in figure 2.1. Another limitation

The effects on abiotic components and their subsequent effects on other components are central to the chain of events. Moreover, not all important effects involve populations; some are economic or aesthetic. Even some populational effects, such as reduction in the trout population, would be overlooked because they are not the direct result of population interactions. However, they *are* the direct result of interactions between the trout and abiotic components (water and soil in the form of mud). Finally, the flourishing of various populations, such as partridges, depends on the farmer *not* engaging in certain types of negative interactions. So, negative interactions (or the lack thereof) are an important part of the story too. Thus, positive and negative effects on organisms, populations, and abiotic components are all relevant to interdependence.

Note that Leopold also emphasized that the health of the land as a whole was something that could be benefited or harmed.[5] According to Leopold, "Health expresses the cooperation of the interdependent parts: soil, water, plants, animals, and people; it implies collective self-renewal and collective self-maintenance" (Leopold 1942b, 300). Again, recent studies have shown Leopold's prescience (or perhaps it would be better to say a remarkable insight) on this point. For example, reintroducing wolves to Yellowstone has arguably produced not only a variety of positive effects on other species, but also changes that "appear to represent the early stage of a recovering ecosystem"; further changes resulting from wolf reintroduction "could represent an important improvement in food resources and physical habitat for an array of wildlife species" and "could also help improve [Yellowstone's] resiliency relative to any ongoing or impending changes in climate" (Beschta and Ripple 2012, 137).[6] Thus, effects on land communities, considered holistically, are part of interdependence as well.

---

is that it does not illustrate the conflicts of interest that can manifest in a land community; these are discussed in subsequent chapters, particularly chapter 6.

5. See chapter 4 for a discussion of *land health*.

6. Here I should acknowledge that Beschta and Ripple's claims have been subject to criticism, but it is equally important to note that they have responded to those criticisms in subsequent publications. The role of wolves in Yellowstone is an active area of research. As I noted in chapter 1, at various points in this book I cite empirical claims that are controversial but that I am not in a position to resolve. This is, unfortunately, unavoidable, but here as elsewhere, these examples are in any case simply meant to be illustrative.

Finally, although I have emphasized in this section that whether a causal interaction is positive or negative in this context depends on the types of effects on the entities involved rather than the types of causes, it is also worth considering which entities are involved in the interactions. Clearly and unproblematically, there are interactions between organisms of the same species and of different species, both direct and indirect (i.e., mediated through other biotic and abiotic components).[7] Perhaps more metaphysically challenging is the question whether populations of different species can interact qua populations. Attempting to answer that metaphysical question would take this discussion off on a large tangent (see Millstein 2013 for a defense of the idea), but I can make a few suggestive remarks toward the idea of populations interacting. Consider again the wolves reintroduced to Yellowstone. Laundré et al. (2010) argue that the amount, location, and types of plant consumed by the elk in Yellowstone has changed in the presence of the wolves. They further suggest that this is due to a fear of the risk of predation (a fear that, they argue, is present in many predator-prey interactions). Arguably, however, fear-inducing risk is present only when the wolves are numerous enough to be a significant risk. Thus, the population size of the wolves and their behavior as a whole jointly change the behavior of the elk as a whole. Competitive interactions mediated by abiotic components are likewise plausibly populational interactions when there is a limited resource; for example, if a population of one plant species sprouts before another, the first can often crowd out the second.

CONSIDERING AN INTERIM ACCOUNT OF INTERDEPENDENCE

Thus, for a complete picture of the connections within a land community, interdependence needs to include different types of positive *and* negative effects and effects on more than just populations, including effects on abiotic components. It also needs to include humans, an essential part of Leopold's story. In short,

---

7. Direct and indirect interactions are discussed further in the section headed "A Web of Interdependencies."

interdependence is broader than the list of typical ecological interactions found in many textbooks (presented above under "Does Interdependence Result Only from Positive Interactions?"), and it includes causal interactions involving humans. However, there are at least two possible objections to this Leopoldian concept of interdependence.

Objection 1: The Leopoldian Conception of Interdependence Is Too Strong

One might worry that Leopold's claim for unity is too strong, given our contemporary understanding of the causal interconnections between biotic and abiotic entities on this planet. Is it really correct to say, as Leopold does, that "all parts of the land, including ourselves, prosper or decline together"? Here it might seem as though Leopold is claiming something akin to what Jay Odenbaugh has recently dubbed as the "mantra" that "everything is connected to everything else" (Commoner 1971), which, Odenbaugh notes, seems to commit us to there being "simply one thing, the universe" (Odenbaugh 2010, 241).

In reply to this objection, it's important to recognize that Leopold need not be committed to this "mantra" or to there being only one planetary ecosystem ("Gaia" or the like). Note that his example of the Wisconsin farmer is not a global one. Rather, his point may simply be that interdependencies can be more extensive than we often realize, so that our fates are in fact tied to entities that we might not typically see ourselves as connected to. As Leopold stated, "There are two spiritual dangers in not owning a farm. One is the danger of supposing that breakfast comes from the grocery, and the other that heat comes from the furnace" (1949, 12). Some people might not think that farming (or mining) practices significantly affect them; Leopold's example shows that such thinking is mistaken.

It is also worth noting that the causal interactions that give rise to interdependencies do vary in strength, and that those variations in strength can be used to circumscribe entities smaller than the universe or the planet: that is, land communities.[8] Thus, one way

---

8. Land communities are discussed in chapter 3.

of understanding Leopold's point might be to say that all parts of a land community, including ourselves, prosper or decline together. Of course, Leopold readily acknowledged that some land communities can adjust to large alterations (e.g., as found in Western Europe) even as others (e.g., in the southwestern and midwestern United States—recall the 1930s Dust Bowl, which Leopold lived through) cannot.[9] The point is that because of the causal interactions between abiotic and biotic components, changes to one part of a land community will cause changes in another, although such changes may sometimes be small. Leopoldian interdependence is completely consistent with some interactions being weak, although Leopold insisted that our judgments of such things are often mistaken (Leopold 1949). It is also worth emphasizing that, as I suggest below, it is really the entire web of interdependencies that matters, so that focusing on a particular interaction as "weak" may be missing the bigger picture.[10]

Those who hold one of the two extreme views—either "everything is strongly connected to everything else" or "all connections between members of the land community are weak"—may not be fully persuaded by Leopold's land ethic as I have characterized it. As with other empirical matters, this issue cannot be settled by this book, so here I note only that all the land ethic requires is that the nature of interactions fall between these two extremes. I think this is a fairly common and plausible view, but readers will ultimately have to decide that for themselves.

Objection 2: Not All Members of the Land Community "Need" Each Other

Still, there might be linguistic resistance to including negative causal interactions as "dependencies" if dependence is understood as "need." In particular, many organisms seem not to "need"

---

9. This "adjustment" hints at an evolutionary dimension to Leopold's concept of interdependence, as does his inclusion of competition and the feedback between biotic and abiotic components (see Millstein 2020b).

10. In addition, there is some reason to think that weak interactions can have significant effects on community stability (McCann et al. 1998, Kadoya and McCann 2015).

humans—if anything, the opposite seems true. In response, it is important to recognize that "need" is only one way of understanding "dependence"; "dependence" can also be understood as "vulnerability" (Anstett et al. 1997).[11] Organisms can be harmed in many ways by a variety of causes and are therefore vulnerable. A vulnerable organism depends on other entities *not* to cause harm to it. With vulnerability, it is easier to see why it is natural to include negative as well as positive causal interactions as part of interdependence.

The entities in Leopold's story of the Wisconsin farmer are all vulnerable and therefore all depend on each other. And organisms do "need" us *not* to do things that will negatively affect them. Like any other organism, humans are vulnerable: the ocean-mediated effects of human-caused climate change on human (and nonhuman) island populations are an obvious example.

This understanding of *dependence as vulnerability* addresses a potential concern with exploitative interactions, mentioned briefly above under "Does Interdependence Result Only from Positive Interactions?" The destruction of habitat for commercial development is a typical example. Many exploitations might be thought to have such dramatically negative effects that, unlike predation or parasitism, there could be no positive impact on other members of the land community. Presuming such cases exist (and I think they do), it might be hard to see them as interactions that underlie interdependence. But if we recognize that interdependence includes the vulnerability of other members of the land community, then these members are dependent on humans to avoid engaging in certain types of activities. The situation is perhaps akin to a doctor's oath to "do no harm"—not in the sense that we can avoid harming other community members entirely (we can't), but in the sense that patients depend on their doctors to do their best to follow that guideline.

Yet, I have emphasized throughout that interdependencies exist not only between biotic entities but between abiotic entities as well—and it might seem strained, at best, to say that abiotic components of a land community are "vulnerable." There are at least

---

11. As far as I know, vulnerability has not been much discussed in the environmental ethics literature, but it has appeared in the environmental humanities literature. See, for example Ginn et al. 2014 and other articles in the same special section of *Environmental Humanities*. Thanks to Marion Hourdequin for this suggestion.

two possible responses. One would be to make the case that abiotic components can indeed be vulnerable. For example, soil can be washed away or depleted of nutrients, air can become polluted, and water can become congested or acidic, and these effects result from interactions between them and other components of the land community. These examples may be seen not as harms to the abiotic components themselves; they may only be seen as harms insofar as they harm biotic components, with "unhealthful" soil, air, and water being unhealthful just for the biotic organisms that depend on them or for the land community as a whole. I don't think it needs to be decided whether abiotic components can be literally harmed or only metaphorically harmed, with their harms serving as a proxy for harms to biotic components and their network of interconnections. Either abiotic components are literally vulnerable or they are only metaphorically vulnerable—both ways of understanding express the interdependency.

Here it is worth noting that the use of the term *abiotic* is mine—one that I use as a shorthand in speaking to a contemporary audience—and not Leopold's, who goes so far as to include soil and water among the biota (Leopold 1939a). If *biotic* merely refers to things that relate to or result from living things, then soil, air, and water, all of which are deeply affected by living things, are clearly biotic. It would be an additional step to say that these entities can be alive, but if that step can be taken, vulnerability can be easily attributed to them.

But should this line of argument still prove unpersuasive, one could abandon vulnerability altogether and understand "dependence" simply as "causal dependence," meaning just that B is dependent on A because changes in A produce changes in B.[12] "Interdependence" involves at least this if not something more. There are causal interactions between components of the land community such that changes in one component produce changes in another. Here, abiotic components pose no special problem. For example,

---

12. Thanks to Rick Morris for this suggestion. Note that in using the phrase "causal dependence," I don't mean to invoke the "causal dependence" accounts of causation discussed in the philosophical literature specifically. The arguments in this chapter are meant to be noncommittal with respect to philosophical accounts of causation.

soils can provide the nutrients that allow certain types of plants to flourish; dead and decaying plants furnish nutrients back to the soil. Nutrient-poor soil, on the other hand, will not promote the flourishing of plants, and certain plants will deplete soils of more nutrients than they return. Of course, not all relationships are reciprocal in this way,[13] which raises the question whether we should be thinking not just of pair-wise interactions, but also entire webs, or networks, of interacting components: a web of interdependencies. Indeed, Leopold thought so.

A WEB OF INTERDEPENDENCIES

Leopold stated that "the new science of ecology . . . is daily uncovering a web of interdependencies so intricate as to amaze—were he here—even Darwin himself, who, of all men, should have the least cause to tremble before the veil" (unpublished note from Leopold, quoted in Meine 2010, 359). Consider yet again the story of the Wisconsin farmer, which shows that it would be a mistake to take a binary (or even trinary) approach toward understanding interdependence. Rather, one needs to consider the whole network, or web, of causal interactions, some of which are *direct* and some of which are *indirect*. For example, the farmer doesn't have a direct interaction with trout, but through a chain of causal interactions, he may have an indirect effect on trout. This makes the trout dependent on the farmer. Furthermore, it may be that not all of the direct causal relationships are bidirectional; nevertheless, they are all part of the same network of causes.

Or consider the following organisms and their habitat in central coastal California: "The fast-growing population of otters . . . has revitalized the eelgrass beds in the once-degraded waterway, which meanders from the headwaters in San Benito County through Moss Landing and flows out into Monterey Bay. The otters eat the crabs

---

13. That being said, I think there are interesting parallels to be drawn between Robin Kimmerer's (2013) conception of reciprocity and Leopold's conception of interdependence, especially in terms of the obligations that arise from interdependence and the benefits that humans are already receiving from other members of the land community, although Whyte (2015) urges caution in making such abstract comparisons.

that feed on the sea slugs that consume the algae that kill the eelgrass. With fewer crabs ... the sea slugs proliferated and devoured algae, allowing the eelgrass to flourish. That, in turn, has reduced mud and erosion in the tidal creeks and channels, revived fish and invertebrate populations and increased nutrients in the estuary" (Fimrite 2018).

This chain of interactions, yielding some direct and some indirect interdependencies, including abiotic components, echoes the web of interdependencies in the Wisconsin farmer story. And although humans are not mentioned explicitly, it is not hard to see various points at which they might interact (e.g., via fishing). As Leopold reflected, "Who knows for what purpose cranes and condors, otters and grizzlies may some day be used?" (Leopold 1949, 220).

Here it is also worth noting that Leopold (1942a; 1949) also spoke of the land community as forming a *circuit*; typically, he was referring to the circuit from soil and back to soil, but the Wisconsin farmer example begins and ends with the farm, so it is generalizable. These considerations point to the desirability of paying attention to networks of interdependence.

Indeed, network thinking has become common in ecology and evolution (Proulx et al. 2005). For example, ecologists such as Valiente-Banuet et al. (2014) have found that network analysis has benefits: it allows testing for nonrandom patterns in interactions, and it is useful for extracting characteristics such as the number of different species with which a certain species interacts or nested patterns of interaction. Other ecologists point out the limitations of pair-wise analyses between pairs (and even multiple pairs) of species populations, due to phenomena such as the presence of genetic correlations between traits involved in multiple interactions or interactions with one species that alter the likelihood or intensity of interactions with other species (Strauss and Irwin 2004), which dictates the necessity of other methods.

A PROPOSAL FOR UNDERSTANDING LEOPOLDIAN INTERDEPENDENCE

The previous two sections lead to an expansion of the interim Leopoldian account of interdependence I gave earlier in this chapter: namely, *Leopoldian interdependence* consists of direct and indirect "negative" and "positive" causal interactions between organisms (including humans), populations, and abiotic components ("interactors") that yield a variety of needs and vulnerabilities in organisms, populations, and abiotic components (as well as land communities more holistically), with interactions that vary in strength and direction in time and in place, often forming a web or network of such interactions.

Let us now revisit the case of the Kaibab presented at the outset of this chapter in light of this elaborated understanding of interdependence. The different elements of this conception can guide us in seeking out relevant aspects of the case that might otherwise be overlooked—one of the benefits of a philosophical elaboration of a central concept.

We have already discussed how individual deer and deer populations can be affected both positively and negatively in their interactions with wolves and plants. And clearly wolves, deer, and plants are all vulnerable in light of their interactions, with plants suffering from too many deer and wolves suffering from too few deer. Humans need to be recognized as part of the web too, since, according to Leopold, the deliberate killing of wolves by humans between 1910 and 1925 was the primary reason there were no wolves on the Kaibab (Leopold 1943). Leopold cites other human interactions as well, such as fire control and human hunting of deer.

But there are far more species in the web than that. The term *plants*, of course, covers a variety of species; Leopold stated that "on the Kaibab, deer pressure was first visible on cliffrose. As this good food became scarce, juniper and finally piñon pine were taken, and fawns began to die" (Leopold 1943, 359). As for predators, it wasn't just wolves and humans that were killing deer; Leopold mentioned other predators, such as cougars and coyote (also killed by humans in turn). Binkley et al. (2006) identify other potential interactors and interactions: aspen, climate, sheep and cattle (present due to humans, of course), human logging, and rodents.

Continuing to use the characterization of interdependence as a guide, once we have identified as many of the interactors and direct and indirect interactions as we can (including more information about the role of abiotic factors such as soil and water), we need to determine in what circumstances and in what ways these interactions are positive or negative and the strength of different interactions under different circumstances. It should be clear that this could quickly get very complex. As is often the case in science, the richer our picture, the harder it is to work with and comprehend, but the more factors we leave out, the worse our predictions of future states of the system (in this case, component populations or the web/network itself) may be, depending on the significance of the factors left out. Leopold would also remind us that there will always be interactors and interactions that we do not know about, so that we should approach our predictions with caution and a healthy degree of skepticism.

In any case, the point here is that the simple wolf-deer-plant picture described at the outset of the chapter is only part of the interdependence story of the Kaibab. It may be a useful pedagogical tool, but Leopold described a more complex picture than that, and rightly so.

CONCLUSION

In this chapter I have argued that Leopold's writings in light of coherence considerations as well as contemporary ecological findings suggest the concept of interdependence described in the preceding section. Given Leopold's influence across ecology and related fields (conservation biology, forestry, wildlife management, restoration ecology, etc.), all of which frequently use the concept of interdependence, this analysis can potentially help to provide a schema that can be filled out in particular cases, taking care to consider all the elements of the concept (biotic and abiotic components, positive and negative interactions, strength of interactions, etc.)

Subsequent chapters illuminate the central role that interdependence plays in Leopold's land ethic. It is crucial, for example, to Leopold's understanding of land health, and it serves as the basis for our obligations to land communities. Meanwhile, the reader may have wondered about the connection between a web of interdependencies and a land community. The next chapter clarifies that connection and offers an account of land communities.

CHAPTER THREE

Land Communities

Conservation becomes possible only when man assumes the role of citizen in a community of which soils and waters, plants and animals are fellow members, each dependent on the others, and each entitled to his place in the sun.

ALDO LEOPOLD, "Original Foreword to *A Sand County Almanac*"

Parts of this chapter originally appeared in Roberta L. Millstein, "Is Aldo Leopold's 'Land Community' an Individual?," in *Individuation, Process, and Scientific Practices*, edited by O. Bueno, R. Chen, and M. B. Fagan (Oxford: Oxford University Press, 2018), 279–302.

INTRODUCTION: MIGRATORY GEESE

The previous chapter ended by noting that interdependencies between biotic and abiotic entities often form a web—an intricate web, according to Leopold. It also hinted at the challenges of deciding which interactors and interactions should be considered part of the web. The epigraph to this chapter suggests that Leopold saw a link between the idea of interdependence and the idea of community. Thus, another way of putting these challenges, in Leopoldian terms, would be to ask, "How should we understand communities?" How do we know which entities and processes are included in a given community and which are not?

Consider the Bosque del Apache National Wildlife Refuge (BdANWR), located in the southern portion of the middle Rio Grande Valley, about 156 km south of Albuquerque, New Mexico.[1] The BdANWR currently has twenty-four wetlands totaling about 494 hectares along the west bank of the Rio Grande. These wetlands are managed by humans and supplied with water through a complex system of canals and drains. They are home to many organisms, including species of algae, water fleas, cranes (some endangered), mosquitofish (*Gambusia affinis*), and crayfish (*Procambarus* and *Cambarus* spp.), as well as various species from the Rio Grande and invertebrates characteristic of pond environments in the southwestern United States. The boundaries of these wetlands are quite visible; they are sufficiently well defined as to be mappable. So, it might be tempting to say that the boundaries of the communities

---

1. Discussion of the BdANWR in this section is drawn from Post et al. 1998; Kitchell et al. 1999; and Post et al. 2007.

are the same as the boundaries of each of the wetlands, and that all of the interdependent species within those boundaries constitute the community—but there are complications.

Every year, over forty thousand lesser snow geese (*Chen caerulescens caerulescens*) and Ross's geese (*Chen rossii*) migrate south (the snow geese come from as far away as Alaska) and spend the winter in the BdANWR. Furthermore, even within the refuge, the geese do not spend all of their time in the wetlands. Once or twice a day, they make an approximately 12 km round trip to agricultural fields (largely corn and alfalfa, managed specifically for wintering waterfowl), returning primarily to one preferred wetland area, identified as Pond 18d by Kitchell et al. (1999). Researchers have estimated that through their guano, the geese supply nearly 40 percent of the nitrogen and 75 percent of the phosphorus entering Pond 18d. This may promote the destruction of wetland vegetation, impose heavy losses on local agricultural crops, increase the risk of infectious disease outbreaks, and decrease water quality. The geese have the potential to have considerable impacts on biotic and abiotic components of Pond 18d, and clearly the geese depend on biotic and abiotic components of that pond. In other words, they are interdependent, at least in the winter season. So, without getting into the details of what a community is (discussed below), there seems to be a reasonable case for saying the geese are part of the Pond 18d community, at least if the time scale covers the winter months. But what about the agricultural fields that the geese feed on daily during the winter months? What about the other wetlands? What about the Rio Grande? What about the northern areas from whence the geese make their annual southern migration? Are any or all of these part of the community, too?

It will take some time to answer these questions. In order to do that, we first need to delve more deeply into Leopold's understanding of community in ecological contexts. I begin with some of the arguments against Leopold's conception of community. I then give a broad characterization of Leopold's conception, noting the ways in which it is similar to and different from other approaches. This characterization, however, does not address the biggest challenge to Leopold's community concept: namely, whether and how boundaries can be drawn—the "Who is in the community?" question. Subsequent sections outline this problem and show how it can be addressed in a Leopoldian manner. I then offer some applications and discussion (including a brief discussion of scales larger than the land community scale).

LAND COMMUNITIES AS A FOCUS OF LEOPOLD'S LAND ETHIC

In the essay "The Land Ethic," Leopold famously stated, "A thing is right when it tends to preserve the integrity, stability, and beauty of the *biotic community*. It is wrong when it tends otherwise" (Leopold 1949, 224–25; emphasis added). This passage and Leopold's land ethic more generally imply that the biotic community is a locus of direct moral obligation or even, some argue, an entity with intrinsic value (Callicott 1987; 2014). But what did Leopold mean by "biotic community"? Interestingly, Leopold considered the biotic community to include not only biotic components but also abiotic components: "soils, waters, plants, and animals, or collectively: the land" (Leopold 1949, 204), which is consistent with the understanding of interdependence explained in the previous chapter. So, *biotic community* is a potentially misleading term; *land community*, a term that Leopold employs in "The Land Ethic" interchangeably with *biotic community*, seems more appropriate—and I use it for the remainder of this chapter. (That being said, if—as noted in the previous chapter—Leopold considered soils and waters to be biotic, then the term is fully appropriate, albeit still potentially misleading to contemporary ears).

Regardless of what it is called, some authors have raised concerns about making the land community the ethical locus of the land ethic. Kristin Shrader-Frechette sums up these worries well:

> Nor is it obvious how to define the system at issue. The ecological problem of defining the system at issue is analogous to the economic problem of defining a theory of social choice and choosing some "whole" that aggregates or represents numerous individual choices. Defining an ecological "whole" to which Callicott and Leopold can refer is especially problematic, both because the biologists (e.g., Clements, Elton, Forbes) cited by Callicott to explicate his [Leopold's] views are no longer accepted by most contemporary scientists as having correct views about ecological communities, and because the contemporary variant of Clements's position,

the GAIA hypothesis, has been rejected by most ecologists as an unproved metaphor or mere speculation. At best it is an hypothesis. They admit the scientific facts of interconnectedness and coevolution on a small scale, but they point out that particular ecosystems and communities do not *persist* through time. Hence, there is no clear referent for the alleged "dynamic stability" of an ecosystem or community. (Shrader-Frechette 1996, 60)

From this passage, I glean the following concerns:

1. It is not clear how to define a land community, which Shrader-Frechette seems to think of in terms of an ecological community or an ecosystem.
2. The concept of "land community," at least as explicated by J. Baird Callicott in his earlier work, is outdated and rejected by contemporary science, and the closest contemporary view has also been rejected by most contemporary scientists.
3. Particular land communities are not things that persist through long periods of time.
4. Thus, there is no clear referent for the land ethic and the stability that it seeks to promote.

Indeed, even Callicott, who has been called the "leading philosophical exponent of Aldo Leopold's land ethic" (Norton 2002, 127), thinks that particular land communities cannot be clearly identified, adding to Shrader-Frechette's list of concerns the following:

5. The boundaries of communities and ecosystems are not fixed by nature, but rather determined by the scientific questions that ecologists pose. (The view that boundaries are not fixed by nature is defended by Callicott [2014] as well as Eliot [2013]).

Some of these concerns involve the idea that the land community is not a distinct and cohesive entity, a biological *whole*, in the way that an organism is usually understood to be a biological whole.[2] The theme of communities as organisms appears in sev-

---

2. Elsewhere I characterize land communities as "individuals" (Millstein 2018) rather than wholes, drawing inspiration from the Ghiselin-Hull thesis that species are individuals (Ghiselin 1974, 1997; Hull 1976, 1978). However, for the purposes of this chapter, nothing depends on the particular philosophical (or metaphysical) claim that land communities are individuals. Instead, here I preferentially use Leopold's own language.

eral of Leopold's essays, and figured in his speech at the dedication ceremony for the University of Wisconsin Arboretum and Wild Life Refuge in 1934: "Ecology tells us that no animal—not even man—can be regarded as independent of his environment. Plants, animals, men, and soil are a community of interdependent parts, an organism. No organism can survive the decadence of a member. Mr. Babbitt is no more a separate entity than is his left arm or a single cell of his biceps" (Leopold 1934; see also 1935b; 1944a).

Relatedly, Leopold "sketch[ed] the concept of land-as-a-whole" (Leopold 1944a, 310), described land health (or land illness) as an attribute of the community as a whole, and suggested that "the components of land have a collective as well as separate welfare" (Leopold 1944a, 316). By the end of his life, he seemed to de-emphasize (although not eliminate) his characterization of communities as organisms, but even then, he still maintained that the members of communities are interdependent; referred to soils, waters, plants, and animals *collectively*; and spoke of the organization (or, in degraded situations, the disorganization) of land (Leopold 1949).

But can these views of Leopold's—seeing land communities as akin to organisms or biological wholes—be defended, or is the land ethic undercut because the land community concept is too problematic, as Shrader-Frechette suggests? My ultimate goal is to see if there is a *defensible* concept of a land community as a biological whole that is close to Leopold's stated views. I argue that there is. But rather than looking only to Clements, Elton, and others to understand Leopold's conception of land community, as Callicott and Shrader-Frechette seem to do, let us begin with Leopold himself.

LAND COMMUNITIES AS BLENDED COMMUNITY-ECOSYSTEMS

Land, Leopold tells us, "is not merely soil; it is a fountain of energy flowing through a circuit of soils, plants, and animals. Food chains are the living channels which conduct energy upward; death and decay return it to the soil" (1949, 216). So, Leopold's concept of a land community not only included abiotic components, as mentioned above, but was also at least partially characterized in terms of *matter and energy flow*. This was represented by a biotic/land pyramid (visually depicted in a diagram in his 1939 essay, "A Biotic View of Land"; see fig. 2.2) that he described as follows: "Plants absorb energy from the sun. This energy flows through a circuit called the biota, which may be represented by a pyramid consisting of layers. The bottom layer is the soil. A plant layer rests on the soil, an insect layer on the plants, a bird and rodent layer on the insects, and so on up through various animal groups to the apex layer, which consists of the larger carnivores" (Leopold 1949, 215).

However, as previously noted, Leopold also emphasized that a land community is composed of *interdependent parts*. Drawing on the work of community ecologist Charles Elton (1927), Leopold described a complex tangle of lines of dependency for food and other "services" such as shade (see fig. 2.1).[3] Food chains are sometimes thought of as just an energy conduit, but for Leopold they also represented trophic (feeding) relationships and other types of relationships between members of different species. Leopold thus stressed the interactions between organisms and the way in which changes in some species affect other species, and he did so throughout "The Land Ethic" and elsewhere. For example, in "Deer Irruptions," he analyzed the consequences of deer overbrowsing in the absence of predators (Leopold 1943).

Leopold's land community concept thus emphasizes matter and energy flow through organisms and abiotic components and *also* emphasizes the interdependence among organisms and abiotic components. Applying this concept to the example of the migratory geese at the beginning of this chapter, the land community could include the flow of matter and energy from agricultural soil through corn and alfalfa, and through geese to the pond and its inhabitants *as well as* the interactions among biotic and abiotic components, such as the water-dwelling mosquitofish's consumption of water fleas. The incorporation of both aspects is notable because some authors (e.g., Callicott and Mumford 1997; Odenbaugh 2007)

---

3. The ideas of food chains, land pyramid, and community can all be traced back to Elton; see Warren 2013 for further discussion. The idea of interdependence also has some Eltonian roots, but its Darwinian roots may be stronger (see Millstein 2015).

have differentiated between two types of entities and two types of approaches in ecology: ecological communities with a community ecology approach on the one hand, and ecosystems with an ecosystem ecology approach on the other. It is the latter approach, the ecosystem approach, that includes abiotic components and invokes matter and energy flow, de-emphasizing (or even disregarding) organisms and populations. This approach contrasts with the former approach, the community ecology approach, which emphasizes interactions between organisms. Therefore, Leopold's land community concept combines aspects of the concept of ecological community, as it is typically conceived, with aspects of the concept of ecosystem, as it is typically conceived.

One might worry that Leopold's concept of a land community is too idiosyncratic. However, I have argued elsewhere that he was both influenced by and an influencer of a number of prominent ecologists and that, moreover, there are *contemporary* candidates for a land community concept that combines community and ecosystem elements (Millstein 2018). Various authors, such as O'Neill (2001); Chapin et al. (2011); Schulze et al. (2005); and Hastings and Gross (2012), all combine ecological community and ecosystem elements (again, interactions between organisms and matter/energy flow, respectively) in describing the entities that they study. Thus, we can set aside the second concern in the list provided in the second section of this chapter: Leopold's conception of land community is neither rejected nor outdated.

From the contemporary ecologists who combine community and ecosystem approaches, we can draw some important and useful insights (see Millstein 2018 for further discussion). First, the approach that considers matter/energy flow alone creates problems, as does the approach that considers only population interactions (O'Neill 2001). Second, sustainability, rather than stability, may be the relevant property of community-ecosystem entities (see O'Neill 2001). Chapin et al. (2011) also explicitly advocate a non-equilibrium approach. Third, what an ecosystem *is* may be different from its *processes* (Chapin et al. 2011) and its *models* (Hastings and Gross 2012). Fourth, some purported boundaries may exclude relevant processes and thus be inappropriate (Schulze et al. 2005). Fifth, these ecologists collectively challenge the picture of community ecology and ecosystem ecology as distinct approaches.

But is this combination tenable? Does it complicate the case for the land community as a distinct whole?

THE PROBLEM OF BOUNDARIES FOR A LAND COMMUNITY CONCEPT

The question of boundaries could cause problems for thinking about a combined community-ecosystem entity as a biological whole. When we say that an organism is a biological whole, we typically think it can be distinguished from other such organisms. However, without being able to identify boundaries (at least fuzzy boundaries), then the idea of a distinct biological whole seems to dissipate. Note that boundaries in the sense meant here need not be physical, visually inspectable boundaries. Indeed, the Leopoldian view of boundaries that I defend in this chapter is based on interactions, not on physical boundaries; as a consequence, land communities are not necessarily contiguous in space, just as a university can be discontiguous even while remaining one entity, or so I will argue.

This potential concern for a combined community-ecosystem approach is recognized even by those who endorse such an approach. Post et al. (2007) characterize the problem as follows. Usually, they maintain, we can fairly readily see that boundaries are set by discontinuities or steep gradients in the flux and flow of materials and energy and/or by discontinuities or steep gradients in interactions between populations of different species. However, they point out that, whereas some systems are *well bounded* (which is not to say "closed"), others are *open*. In well-bounded systems (e.g., lakes, islands) these two criteria coincide—and coincide with physical boundaries as well—making delineating ecosystem boundaries relatively straightforward. In such systems, "interactions among organisms are typically stronger and cycling of materials and energy is typically tighter within than across the physical boundaries of these ecosystems" (115). On the other hand, in open systems (e.g., most terrestrial habitats, estuaries, and streams), the two approaches do not coincide, as when resources come from areas where species are not interacting (e.g., upstream).

The problem, then, is how to understand the boundaries of open systems. Unfortunately, to my knowledge, Leopold does not address the issue of land community boundaries.[4] So again, we will have to fill in a bit with contemporary work and then see if his basic idea is defensible. Post et al. (2007) describe various challenging scenarios for understanding the boundaries of open systems. Recall the migratory geese scenario, in which large "inputs" (large amounts of phosphorus and nitrogen from geese guano) are coming from the "outside" (in this case, outside Pond 18d) at short temporal scales (once or twice a day). Other animals that migrate or move to lakes, wetlands, and stream can produce similar situations, potentially moving large amounts of nutrients around the landscape to what might otherwise seem like closed systems. In such cases, Post and colleagues suggest, we should recognize that the system is larger than we might have initially thought. We would have to recognize not only the geese as part of the system but also the sources of their food: the agricultural lands that are approximately 6 km away.

But suppose the situation were different. Suppose that internal cycling of material/energy were stronger than external inputs: for example, when a watershed is visited by only a few mobile organisms, or when it has a very high productivity. In such cases, Post and colleagues maintain, we should consider internal cycling to dictate the boundary.

However, recall that the geese move around the landscape not just within their southern wintering areas; they also migrate from up north. So, on a longer time scale, the boundaries might be affected still further. Following the solution that Post and colleagues have provided, whether the boundaries were larger would depend on the relative contribution of the northern inputs to the southern system. The longer time scale doesn't necessarily mean a larger system in this case, but it might.

I have used the word *solution* here, but Post and colleagues do not. Instead, they conclude, "In open ecosystems where there is little or no congruence among physical and functional boundaries, . . .

---

4. I'm not sure why he didn't address the issue, and I don't wish to speculate. However, he does presciently ask, "Does the wild goose, reconnoitering the farmer's cornfield, bring something more than wild music from the lake, take something more than waste corn from his field?" (Leopold 1941b, 22), which is at the heart of the question of boundaries that is being addressed.

each different question may dictate very different definitions of ecosystem boundaries" (2007, 122). Although this is an interesting position worth considering, it doesn't seem to me that Post and colleagues have made the case for it. That is, they don't seem to have described situations where different questions would indicate different boundaries; on the contrary, as I describe in the preceding paragraphs, they seem to have offered solutions to such cases where one might think that such a problem arose.

Perhaps their concern is that, in cases like that of the migrating geese, the size of the system or its degree of boundedness can vary depending on the time scale, with, as they note, all systems being open given a long enough time scale. Still, it seems like we could also say that they have offered a scheme for understanding how, *given* a time scale, system boundaries ought to be characterized. Time scale would then be an essential parameter for land community boundaries, perhaps analogous to the way that the size of an organism's boundaries depends on the time scale in question. (Such a shift in size and boundaries might be particularly striking for a long-lived organism like a quaking aspen, where what appears to be a forest is actually one system connected by its root system underground). On the way of thinking that I am suggesting here, it is not that different questions are determining different boundaries per se, but rather that the boundaries actually do change over time.

But even taking the stated conclusion of Post and his colleagues at face value, a question remains: Are open systems where different questions dictate different ecosystem definitions and different ecosystem boundaries coherent enough to be entities that we owe direct obligations to or entities that have intrinsic value? (This is the fifth concern from the list in the second section of this chapter). This question has an ontological component and an ethical component, which I discuss in turn.

In order to address the ontological component—namely, the status of interest-relative entities—it will be instructive to consider similar views. Callicott likewise holds that ecosystems "are in effect defined, both spatially and temporally, by the ecological question posed" (2014, 3), yet he maintains that they are "real, existing entities" (94)—at least in part. That is, he asserts that "when we come to isolate them, to bound them, for purposes of ecological study, we partly create them" similar to the way, he says, that "electrons emerge fully into existence when quantum physicists measure them" (41). Setting aside questions about his interpretation of physics and the strangeness of this analogy, it is difficult to understand what Callicott means. Does he mean that, in the absence of investigators, there really are no ecosystems or communities, at least not in a full sense? Perhaps not, but then it seems as though one cannot continue to defend ecosystems and communities as distinct biological wholes or as "real, existing entities."

Eliot (2013) offers a more sophisticated version of the argument that our interests partially determine whether something is a community. According to Eliot, boundaries are determined by the set of causal relations relevant to some interest; furthermore, a community is "a real object, in so far as its component populations are connected by a particular kind of causal connection" (8). Odenbaugh (2010) similarly claims that different causal relations may pick out different ecosystems, although he does not tie this claim to our human interests. However, we don't seem to have such loose causal relation requirements for other putative biological wholes, such as organisms. For example, in the human body, "circulates blood" does not fully coincide in physical space with "circulates oxygen," yet we think of those causal relations as picking out the same biological whole (the same organism), not two different biological wholes. This is presumably because the system that circulates blood and the system that circulates oxygen are tightly interconnected with other systems from which they are not fully separable. So we should at least consider whether the same is true for putative communities/ecosystems. The conception of land communities that I argue for below includes, by definition, all such tight interconnections. (Of course, weaker, external causal relations will still be present, but that is likewise true of a human organism as well).

Turning to the ethical component of the question, if (contra to what I argued earlier) Callicott's understanding of the ontology is correct—if ecosystems are real yet interest-relative—then it seems that the moral considerability of an ecosystem would be dependent on an ecologist studying it. If the ecosystem loses moral considerability (indeed, ceases to fully *be* an ecosystem) when it is not being studied, then that is a weak notion of moral considerability indeed.

So, Callicott's understanding of the term *ecosystem* as a characterization of Leopold's land community concept is insufficient for his (and our) ethical purposes, making a continued search for a possible workable alternative desirable.

On the other hand, Eliot's and Odenbaugh's alternative approaches toward characterizing communities/ecosystems, which are ultimately grounded in causal relations, also run into problems concerning moral considerability. Here, the problem arises because of the multiplicity of possible boundaries. To be clear, the worry here is not that boundaries may be a bit fuzzy, since surely many real, existing entities have fuzzy boundaries: for example, to be an organism is to constantly lose and gain cells, yet human organisms are surely "real, existing entities." (We thus ought not be surprised when land communities turn out to have fuzzy boundaries as well, as they do). Rather, the problem is one (as highlighted by Russow 1981 for the case of species) of how many entities there are. Considering a given geographical area, do we have one land community, two land communities, or more? Perhaps the lack of a definitive answer to that question is not problematic on its own, but what if different ways of drawing boundaries for a given geographical area cross-cut each other, so that, in some cases, we have one land community as a subset of another, whereas with others, one land community overlaps with another? This makes the object of our moral obligations unclear, especially if we must choose between different purported land communities when our conservation funding is limited. Can we eliminate or add moral obligations simply by asking different sorts of scientific questions? Are all possible ways of drawing boundaries equally legitimate? It would seem that we could draw an infinite number of boundaries for a given geographic area; in that case, do our moral obligations shift with each possible drawing? That idea seems untenable and unworkable.

Moreover, would it be wise to try to treat purported land communities well while failing to consider some of the population interactions or energy flows relevant to their sustainability? I think one would run into practical problems if one did so. To use the analogy of the human body again, it would be akin to trying to benefit one's arm muscles without consideration of one's heart and lungs; eventually, the arms will fail when the body fails from ill health. So we ought to consider more closely whether we can delineate the boundaries of open systems in more systematic and defensible ways.

RESPONSES TO BOUNDARY PROBLEMS

Indeed, there seem to be (at least) three ways of handling the boundary problem concerning open systems—systems where the spatial area of the densely interacting populations is larger than that of the dense matter/energy flow, or vice versa. One possibility is that *the land community exists within the* larger *of the two areas*. In other words, we always "go big"; if the spatial area of the densely interacting populations is larger than that of the dense matter/energy flow, the land community would consist of the area covered by the densely interacting populations, whereas if the area of the dense matter/energy flow is larger than that of the densely interacting populations, the land community would consist of the area covered by the dense matter/energy flow. For example, one land community might consist of all of the areas that the migrating geese cover, plus the Rio Grande and all associated wetlands, and possibly some other areas as well. However, a possible problem with this approach is that we would lose the concept of the ecosystem as a focal level, going beyond locales that lend themselves to concrete study in the field—perhaps to biomes (Currie 2011). This doesn't strike me as a devastating objection—we could just acknowledge that what we study is always a subset of the entity itself—but the objection is worth taking seriously, especially if there are better alternatives.

A second possibility is that *the land community exists within the* smaller *of the two areas*. In other words, we always "go small"; if the spatial area of the densely interacting populations is smaller than that of the dense matter/energy flow, the land community consists of the area covered by the densely interacting populations, whereas if the area of the dense matter/energy flow is smaller than that of the densely interacting populations, the land community consists of the area covered by the dense matter/energy flow. The problem with this proposal is that it might exclude *causally relevant factors* for the future states of populations and abiotic components and thus give a misleading picture that would be subject to error, making this proposal completely untenable, in my view. For example, suppose that we were to say that Pond 18d, excluding the agricultural lands where geese feed, is a land community. According to Kitchell and colleagues,

> Refuge managers are concerned that high levels of nutrient loading may create water-quality conditions detrimental to these wetland pond systems and possibly create a major problem by loading nutrients to the downstream systems. We have no means for evaluating the prospect of their greatest concern—disease outbreaks—but we can offer the conclusion that effects are local and proportional to bird densities. Refuge managers have extensive experience in manipulating bird abundances. Our general results indicate that nutrient loading rates peak early in winter and *are much higher than would be expected from simple knowledge of bird densities because geese are very actively feeding on high nitrogen foods and translocating much of that to the roosting ponds.* (Kitchell et al. 1999, 835; emphasis added; citations deleted)

In other words, predictions concerning the amounts of nitrogen in Pond 18d and its inhabitants might be highly inaccurate if the nitrogen-rich crops that the geese are feeding on were not taken into account. If we are concerned about ongoing water quality in areas like Pond 18d, we need accurate predictions as well as an accurate accounting of the causes in order to take appropriate measures to protect water quality. But "going small" can prevent us from being able to achieve that successfully.

A third possibility is that *the land community includes interactions or matter/energy flows from the larger area if and only if those interactions or matter/energy flows are stronger or larger than those of the smaller area.* This, in essence, seems to be how Post and his colleagues handle cases such as those where mobile organisms bring large amounts of nutrients into lakes, wetlands, and streams, as I described above. This is the solution that I am inclined to accept, despite the hesitations that Post and colleagues seem to have about it; it promises to preserve land communities as objects of study while taking into account most of the important causal processes that affect the land community's future ("most of" because, since no biological system is closed, there is always the potential for a rare but strong causal influence from the outside).

But there is a potential concern with the third option for characterizing land community boundaries. It might be objected that we do not even have sufficient geographical congruence to constitute ecological communities and ecosystems, much less the congruence between ecological communities and ecosystems that we'd need for land communities. Kim Sterelny, for example, raises concerns about the lack of congruence within purported ecological communities, using Black Mountain (a bush reserve near Australian National University) as an example: "'Black Mountain' names a quite heterogeneous region of about 10 square kilometers with gentle variation from patch to patch. As a consequence of these gradual changes in character, the different populations might not be correlated. A local brushtail possum population may overlap with a local ringtail possum population, a local boobook owl population, a greater glider population, and a number of eucalyptus populations. For on Black Mountain, there are no sharp changes that matter to all of these species, keeping local populations congruent with one another" (Sterelny 2006, 225).

However, this lack of congruence does not invalidate the third possible solution to the problem of boundaries. It is not necessary that all the populations of a community be located in the same place; what matters is the continuity of causal interaction across the populations, even if, for example, the local ringtail population is not interacting with the local boobook owl population. As long as the interactions among the listed populations are stronger than other, "external" interactions, they are all part of the community. This is analogous to the case of a continuous population (Millstein 2010), where the endpoints of a population spread over space do not interact with each other even though there is interaction among the organisms across the entire space, forming one population.

Moreover, a quick comparison to organisms shows that the parts of a biological whole need not be congruent; just as a heart muscle and a leg muscle are not congruent, the populations (the "parts") of a community need not be congruent. The objection is puzzling.

Sterelny raises a further sort of worry about drawing boundaries for an interactionist account of ecological communities specifically:

> [T]he interaction patterns of different components of putative communities may well not coincide. Even if communities are networks of interacting populations, they are typically demographi-

cally open. Migrants move in and out of most habitat patches, ... [and] such movements are likely to have stabilizing effects. We have two populations rather than one if organisms of the same type are related by metapopulation dynamics rather than competition. The echidnas in Black Mountain are part of a different population, and hence a different community from the echidnas on the O'Connor Ridge (about a kilometer to the north) if they are a source population for the O'Connor Ridge echidnas. They then buffer that group against population collapses rather than competing with them for scarce resources. Prima facie, though, there is not much reason to expect the dynamics of echidna populations to match those of larger and more mobile organisms, or those of smaller and less mobile ones. (Sterelny 2006, 217)

Again, though, I do not see that this is an insurmountable objection. If we have correctly identified the echidnas as forming a metapopulation, then the interactions between those two populations are rare (see Millstein 2010 for a discussion of the metapopulation concept). So, even though these rare interactions may sometimes turn out to be significant (as in the case that Sterelny describes, where one population recolonizes a location where another population has gone extinct or nearly extinct), there is no difficulty in saying that the echidnas in Black Mountain are part of one community and the echidnas on the O'Connor Ridge are part of another. They still represent a situation where there are continuous interactions among Black Mountain populations and among O'Connor Ridge populations with discontinuities between them. Discontinuities do not imply that there are no interactions, only that they are fewer and weaker. And if it were to turn out, contra to supposition, that there were significant migrations and interactions between the echidnas on Black Mountain and the echidnas on the O'Connor Ridge, then we have misidentified the echidnas as a metapopulation; they would instead be a patchy population (see Millstein 2010), and we would then have a case for considering all of the populations (consisting of different species) of O'Connor Ridge and Black Mountain to be one community (since, again, it is not required, as discussed in my response to the first of Sterelny's objections, that every population interact with every other population or that they be in the same place).

I have focused on the objection that the populations within ecological communities lack sufficient congruence for us to identify their boundaries, but the same kinds of responses can be made to those who might claim that the matter/energy flows of ecosystems lack congruence. Recall, however, that the goal of this chapter is not to defend an ecological community concept or an ecosystem concept per se, but to defend a combined ecological community-ecosystem concept—that is, a land community. And I have already described how to address lack of congruence between an ecological community and an ecosystem: the land community includes interactions or matter/energy flows from the larger area if and only if those interactions or matter/energy flows are stronger or larger than those of the smaller area.

A PROPOSAL FOR UNDERSTANDING LEOPOLD'S LAND COMMUNITY CONCEPT

Insights from the preceding sections lead me to propose an elaboration on Leopold's concept of land community:

> A *Leopoldian land community* consists of populations of different species interacting with each other and with their abiotic environment over a specified time scale; these *survival-relevant interactions* often produce a *flow of energy and materials* between biotic components and between biotic components and abiotic components (and vice versa).

Let me expand on this proposal a bit more.

Survival-relevant interactions between the populations include competition for scarce resources, predator-prey, parasite-host, pollinator-pollinated, and provision of shade or shelter—in other words, the interactions discussed in chapter 2 that give rise to interdependencies. *Relevant flows of materials and energy* include primary production (photosynthesis, chemosynthesis), secondary production, evapotranspiration, decomposition, and nutrient cycling. These are not meant to be controversial, or even original, parts of my proposal; they are simply the typical interactions and material/energy flows identified by ecologists. Survival-relevant interactions between the populations can produce flows of materials and energy, but flows of materials and energy can also produce or affect survival-relevant interactions between populations. *Food webs* are of particular importance to a combined community-ecosystem approach because they can represent species interactions within a community *and* energy flow through those species (Post et al. 2007); they are thus of particular importance to a land community. Note that this is not meant to be an exhaustive list of interactions and flows of material/energy, as is discussed further below.

Land community boundaries for *well-bounded systems* are located where discontinuities or steep gradients in the flow of materials and energy coincide with discontinuities or steep gradients in species interactions. Land community boundaries for *open systems* are delineated (at a minimum) by the smaller of the two types of discontinuities or steep gradients, including the more extensive interactions or matter/energy flows, *if and only if* those interactions or matter/energy flows are stronger or larger than those of the smaller area. This approach has the advantage of including all significant causally relevant factors for the future states of populations and abiotic components (interdependencies). It may mean that there are fewer land communities than one might have thought; however, I am not sure that this is a problem. Ecologists may reasonably choose to study subsets of land communities (including particular types of interactions or particular matter or energy flows) for various pragmatic reasons, but such choices would not affect what land communities are.

Although here I have drawn on Post et al. (2007) in using discontinuities or steep gradients to identify boundaries, this approach is similar to Simon's (2002) account of "nearly complete decomposability," which I have used elsewhere in characterizing the concept of population (Millstein 2009; 2010). It is different from Simon's approach in focusing on the strength of interactions and flows—where strength can be understood as the size of the effect that changes in one population produce in another population, or the size of the effect on abiotic components—rather than their rate. This difference should not be seen as a crucial one: *differences in the rate of interactions are likewise relevant for boundaries of land communities.*

I have left the time-scale parameter of the land community open, reflecting the idea that what counts as a land community may depend on the time scale chosen, as the migratory geese example illustrates. This can make land communities time-relative (discussed further in the next section). That being said, bear in mind that the composition of species in a land community may change over time, and the populations it contains may evolve over time. What makes it the same entity—the same biological whole—is the continuity of interaction and material/energy flow within the entity through time. This has implications for concerns about instability of communities. Since their members may change, land communities may or may not be stable in the sense of stasis or equilibrium. Rather, sustainability (similar to what Leopold meant by stability or land health, the topic of the next chapter) may be a more pertinent trait or feature of land communities (O'Neill 2001). For example, Leopold (1944c) traced four epochs of changing species within southwestern Wisconsin, but he does not seem to have thought of these as representing different land communities. Rather, he lamented the loss of land health in the face of changing human land practices, not the change of species itself (although he believed that the latter often negatively affected the former while acknowledging many cases in which it did not do so). This addresses the third concern from the list in the second section of this chapter: the concern that particular land communities are not things that persist through long periods of time.

Land communities have beginnings and endings in time just as other biological wholes do—in the case of Pond 18d, this happens every season. There would have to be a complete loss of continuity (a complete absence of interaction and matter/energy flow) for a land community to go "extinct," although land communities that are so degraded as to be virtually extinct may be more common. On the other hand, given the current state of the environment and current societal practices, new land communities might most often be formed by human-caused splitting of an existing land community, analogous to the way that a bacterium divides to form two new bacteria.

This Leopoldian characterization of land communities in terms of interactions and matter/energy flows challenges the so-called Gleasonian picture of communities as mere coincidental assemblages of whatever organisms happen to be located in a particular place at a particular time. Some philosophers (e.g., Regan 1983) cast doubt on the idea that mere "collections" can have moral rights, so this is a salient point. While it is an empirical question as to whether there are causal interactions among populations of different species that affect their survival (and so, an empirical question as to whether there are land communities in the sense I have described) or whether they are mere assemblages, there is plenty of evidence (more than I could reasonably capture here) that such causal interactions exist. Indeed, Eliot (2011) argues persuasively that Gleason (e.g., Gleason 1917) has been interpreted too radically, noting that "every ecologist, including Gleason, recognizes interactions among organisms, including that some require others, to survive" (Eliot 2011, 102). Once one considers, for example, that trophic interactions (which affect both the eaten and the eater) are sufficient, it becomes virtually impossible to deny the existence of survival-relevant interactions. Recent work suggests that these and other interactions are important enough that the order and timing of species immigration during community assembly can affect species composition and abundances (Fukami 2015). On the other hand, if Gleason turns out to have been right that the compositions of species in an area change frequently, nothing I have argued for is challenged here, since (again) it is the causal interactions and matter/energy flows that are essential to the land community's continuity and not any particular composition of species.

Other authors have challenged traditional conceptions of community. For example, Robert Ricklefs has argued that the local community concept is problematic because it overlooks important regional interactions, but given that he argues for an expanded understanding of community that includes the larger region (Ricklefs 2008), I think his view is consistent with, or at least not terribly different from, the Leopoldian conception that I have presented here. Hubbell's neutral theory challenges the assumption that species at the same trophic level will necessarily differ in terms of their demography, but he clarifies that "distributions of the traits and interactions of species between trophic levels require non-neutral models because they fundamentally involve species differences" (Rosindell et al. 2011, 343). Thus, the Leopoldian interaction-based approach argued for here, which does largely involve interactions between different trophic levels, is not challenged by the neutral theory. And while ecologists such as Rapacciuolo et al. (2014) have argued that different species react differently to changes in climate, as noted above, there is nothing in the Leopoldian approach to communities that insists that species composition must remain the same over time, only that they interact in such a way as to affect one another, even if those interactions likewise change over time.[5] In short, "neo-individualist" approaches in ecology do not challenge the land community concept characterized here and may even be congenial to it.

APPLICATIONS AND DISCUSSION

Let us return once again to the case of the migrating geese. Considering the Leopoldian concept of land community outlined in the previous section, if we examine a winter time scale, the land community includes not only Pond 18d with its various abiotic and biotic entities, interactions between populations of different species, and matter/energy flows, but also the agricultural lands at some distance, with their various abiotic and biotic entities, interactions between populations of different species, and matter/energy flows. The geese, of course, connect the two areas with their daily migrations and deposits of significant amounts of guano.

But what about a longer time scale, such as one that encompasses a full year and thus all four seasons? As I suggested in the preceding section, what constitutes the land community at this longer time scale is an empirical question, depending on the strength and size of matter/energy flows between the New Mexico site and the northern regions where the geese live during the other times of the year. Here it is worth noting that Post et al. (1998) and Kitchell et al. (1999) go to considerable lengths to demonstrate the large effects that the geese's feeding patterns have on the smaller time scale; they show that the transmitted nitrogen and phosphorus comes specifically from the geese that feed on the nearby agricultural lands. So the impact from the geese's migration from northern regions cannot simply be assumed one way or another without further study.

---

5. Large-scale migration of species on short time scales as a result of climate change would surely challenge the identity of land communities—that is, the determination of whether a web of interacting species and abiotic components is the same as or different from an earlier one. While such determinations will be challenging, it may simply be that we end up identifying land communities on much shorter time scales than previously, with the identity of land communities over time being much more fluid.

There are other aspects of this case that raise questions. One relevant feature is that the agricultural lands and wetlands are human-managed specifically for waterfowl. The involvement of humans is not in itself problematic; as argued in the previous chapter, Leopoldian interdependence includes human activities; thus, humans are very much a part of land communities. That Leopold meant to include humans as part of land communities is reflected in the epigraph to this chapter as well as his oft-quoted assertion in "The Land Ethic" that "a land ethic changes the role of *Homo sapiens* from conqueror of the land-community to plain member and citizen of it" (1949, 204). So even though this land community might in some respects be considered not fully "natural," it is a land community nonetheless.

Perhaps more challenging is the way that the wetlands are managed. In the fall, water is drawn from the Rio Grande through canals to fill the ponds at the BdANWR; this is the source of the aquatic organisms that are members of the land communities. In the spring, the ponds are drained, and the biomass returns to the river through the effluent canal system of the refuge. Thus, the pond land communities are *ephemeral* ones. Are they any less land communities for being ephemeral? I think the answer to this question must be "no." As noted in the previous section, we expect biological wholes to have beginnings and endings in time. The human management of the refuge is meant to mimic the natural wetlands that would form every winter prior to the changes that humans made to the Rio Grande. Moreover, ephemeral communities are not unusual in many areas, such as in California, where vernal pools are often formed. Ephemeral land communities simply exist on shorter time scales than other land communities. They are not different in kind.

Examining other cases shows that the Leopoldian land community concept can be used as a tool to clarify various situations. For example, consider the two species of riverine tsetse fly that inhabit the Mauhoun river basin in Burkina Faso, discussed by Peck (2009). Peck suggests that the different species have different perspectives that don't always correlate with landscape features that we humans would pick out in characterizing the ecosystem. *Glossina tachinoides* is better able to cross dry treeless regions than the closely related *Glossina palpalis*, with the result that *G. tachinoides* forms one large panmictic breeding population over the whole area while G. *palpalis* forms into separate, more isolated populations. Thus, Peck concludes, from one species' perspective, the landscape is fragmented, while from the other species' perspective, it is not. While this is certainly the case, it is not the full story—it doesn't capture the land community perspective.[6] So what's missing? The studies that Peck draws on (Bouyer et al. 2007; Kone et al. 2011) give some clues as to what other organisms might be members of the land community: for example, the area is being farmed, and cattle are grazing; there are forests along the river; savanna areas border the forests; and there are adjacent river beds. The land community perspective would ask, What are the other members of the land community? What are the interactions between members of the land community, and how strong and frequent are they? What kinds of matter and energy are being exchanged? Where do the discontinuities/steep gradients occur, allowing us to determine boundaries from the perspective of the land community?

Another example is provided by Gounand and colleagues in a literature review together with a quantitative synthetic assessment of cross-ecosystem flows of carbon connecting the major ecosystem types across the globe. They find (among other things) "spatial couplings in which freshwater and unproductive benthic ecosystems receive quantitatively important material exported from terrestrial and pelagic ecosystems, respectively" (Gounand et al. 2018, 3). This is an important finding, but, from a land community perspective, we would want to ask additional questions: for example, How were the boundaries of the ecosystems determined in the first place? What are the species and interactions within those purported ecosystems? What other flows besides those involving carbon are significant? Answering those questions would give us a better handle on which cases involve distinct entities and which do not; for example, the cases in which the flows between "different" ecosystems are so strong that it doesn't make sense to call them different ecosystems.

These two examples might seem deficient because I am asking questions rather than answering them, but my point here is that we

---

6. Peck and Heiss (2020) endorse the land community perspective.

do not always have the empirical data we need to determine when we have distinct land communities and when we don't. Nonetheless, the land community concept can be useful—can serve as a conceptual tool—by suggesting questions that can help us ascertain land communities and their boundaries.

The study from Gounand and colleagues raises a further question: How should we understand the carbon flow between ecosystems that they describe, remembering that there are no genuinely closed systems? Gounand et al. (2018) refer to these as *meta-ecosystems*. The flows between ecosystems are rarer and weaker than the flows within; the larger meta-ecosystem is a biological whole, but much less cohesive than the ecosystems of which it is composed. Similarly, other scholars refer to *meta-communities* (Leibold et al. 2004) composed of two or more communities that have weak interactions. Analogously, with land communities as combined community-ecosystem entities, we should be able to identify *meta–land communities*. These would be larger, less cohesive entities than land communities themselves, made up of two or more land communities whose members interact and/or between which there is matter/energy flow (but again, at lesser strength and slower rate than within the land communities themselves). The land communities within a meta–land community would thus be interdependent as well, albeit to a lesser degree than the members within a land community are.

The possibility of meta–land communities suggests further questions about scale—for example, Can a land community perspective be extended even further, to a global scale? In an unpublished essay from 1923, Leopold considers the question whether the Earth is an organism, suggesting that, even if we don't consider it to be alive, it has "enormously intricate, and interrelated functions among its parts" and that such a view must "admit the interdependent function of the elements" (Leopold 1923, 95). So it seems fair to ask if we might view the entire Earth as a meta–meta–land community (or meta–meta–meta–land community, etc.).

Newman et al. (2017) suggest that at larger scales, interactions *other than* the common ecological ones discussed in the previous chapter (predation, parasitism, competition, etc.) matter most for species composition—at a medium scale, factors like speciation, migration, and extinction; at a continental scale, factors like geology and climate. Because of this, they argue, the land ethic would give short shrift to large-scale environmental concerns. However, what is essential for Leopold's land ethic is *interdependence*, not any particular manifestation of interdependence. To quote Leopold, "All ethics so far evolved rest upon a single premise: that the individual is a member of a community of interdependent parts" (1949, 203).

A recent article examining the likely effects of human-caused climate change on sixty-six bumble bee species illustrates this global interdependence well. The authors conclude that "overall rates of climate change–related extirpation among species greatly exceed those of colonization, contributing to pronounced bumble bee species declines across both Europe and North America with unknown consequences for the provision of ecosystem services" (Soroye et al. 2020, 687–88). So, humans, by their production of greenhouse gases (CO_2, methane, etc.) have been changing the climate in a way that negatively impacts bumblebee species, which in turn disrupts the "ecosystem services" (such as pollination of crops) that humans rely on bumblebees for. The time scales are longer than those for geese bringing guano to a pond, but they are no less ecologically important. Here we must remember that abiotic elements are part of land communities as well, so that human activities that modify the composition of gases in the atmosphere are very much relevant. In short, a land community perspective encompasses local webs of interdependencies, global webs of interdependencies, and everything in between.

CONCLUSION

Leopold's concept of a land community encompassed both interactions between species populations and matter/energy flows. It thus cannot be fully equated with either ecological community or ecosystem concepts, but contains elements of both. This offers a comprehensive approach that will focus our attention on all the relevant aspects of a system—not just matter/energy flows even if species go extinct, and not just the preservation of species numbers even if important matter/energy flows decline. Both are part and parcel of what must be considered when we turn to thinking about the conservation of land communities, as the epigraph to this chapter directs.

It is challenging, but not impossible, to delineate the (always fuzzy) boundaries of land communities. They can be characterized in terms of discontinuities or steep gradients in the flux and flow of materials and energy and/or by discontinuities or steep gradients in interactions between populations of different species, including strong interactions and large inputs that might otherwise appear to come from the outside. The ability to identify boundaries means that we can understand land communities as distinct biological wholes. We can pick out land communities (and meta–land communities, etc.) and see them as entities that persist and change over time, over a variety of time and spatial scales. This makes them *candidates* for direct moral obligation.

But being a biological whole is not sufficient for direct moral obligation. Many philosophers have argued that for an entity to be a candidate for direct moral obligation, the entity must also be capable of being benefited or harmed. The capacity for land health is just such a capacity—that is, arguably, a land community can be benefited by having its health maintained or promoted, and it can be harmed by having its health impaired. The next chapter thus takes up the topic of *land health*. (The Leopoldian argument that land communities have value and *are* owed direct moral obligation is taken up in chapter 5).

CHAPTER FOUR

Land Health

I made a series of vacation trips to the Sierra Madre in Chihuahua, Mexico, in company with my brother Carl, my friend Raymond J. Roark, and my son Starker, by then grown. The Sierra Madre was an almost exact counterpart of my beloved mountains of Arizona and New Mexico, but fear of Indians had kept the Sierra free from ranches and livestock. It was here that I first clearly realized that land is an organism, that *all my life I had seen only sick land, whereas here was a biota still in perfect aboriginal health*. The term "unspoiled wilderness" took on a new meaning.

ALDO LEOPOLD, "Original Foreword to *A Sand County Almanac*" (emphasis added)

INTRODUCTION: LAND SICKNESS IN THE SOUTHWESTERN UNITED STATES

In a number of essays written at various stages of his life, Leopold returned again and again to his experiences in the southwestern part of the United States and across the border in Mexico. As the epigraph to this chapter suggests—an excerpt from the original foreword to A Sand County Almanac—Leopold was struck by the differences between the mountains in the southwestern United States and the Sierra Madre just over the border in Mexico. After a trip to Chihuahua, Mexico in 1937, Leopold described this difference in detail in an essay entitled, "Conservationist in Mexico." The mountainous areas of Chihuahua, Leopold stated, have "a history and terrain so strikingly similar to southern New Mexico and Arizona," yet the former presents a picture of "ecological health," while the latter areas were "badly damaged," despite national forests, national parks, and "other trappings of conservation" (1937, 394).

Chihuahua had "live oak–dotted hills fat with side oats grama," "pine-clad mesas spangled with flowers," and "lazy trout streams burbling along under great sycamores and cottonwoods," whereas on the US side, "the grama is mostly gone, the mesas are spangled with snakeweed, the trout streams are now cobble-bars" (Leopold 1937, 394). The forests of the Chihuahua Sierras burn every few years with "no ill effects, except that the pines are a bit farther apart than ours, reproduction is scarcer, there is less juniper, and there is much less brush" (395). Significantly, Chihuahua watersheds were "intact," whereas on the US side they were "a wreck." What did Leopold think accounted for the difference? In a word: overgrazing. Overgrazing on the US side, lack of overgrazing on the Mexico side. Overgrazing, Leopold thought, leads to erosion, loss of soil, and thus loss of ability to support the same diversity of plant and animal life.

Another difference that Leopold identified was the abundance of certain animal species. In Arizona and New Mexico, Leopold stated, deer ranges tend to be either "overstocked" or nearly empty—the Kaibab, discussed in chapter 2, being not the only such instance, a phenomenon that Leopold described in the southwestern United States and in other areas. In the mountains of Chihuahua, on the other hand, the deer (and also wild turkeys) were "abundant" but "not excessive," lacking in the deer irruptions found in the United States. Notably, mountain lions and wolves were still common; coyotes were not to be found. In the United States, however, coyotes had "invaded." Here Leopold thought that the presence of wolves and mountain lions accounted for the difference: they kept the deer (and coyotes) in check on the Mexico side, while their deliberate extirpation on the US side had led to boom-and-bust deer population cycles and an influx of destructive coyotes.

The differences, Leopold hastened to add, are not the result of the mere presence of humans. "Hundreds of dams" in the Sierra Madre, themselves hundreds of years old, suggest that Indigenous peoples lived in these areas and modified them. Leopold speculated that the purpose of the dams was irrigation for "little fields or food patches," and he wondered what kinds of foods they might have grown that weren't subject to destruction by local animals. As the epigraph to this chapter implies, such practices did not prevent the Sierra Madre from being an "unspoiled wilderness" in "perfect aboriginal health." And as Leopold later noted in "Song of the Gavilan" (the Rio Gavilán runs through the Sierra Madre), "There once were men capable of inhabiting a river without disrupting the harmony of its life" (Leopold 1949, 150).

Here I make a digression that becomes relevant later in the chapter. Note that Leopold's comparative, observational approach to understanding the Sierra Madre and the southwestern United States can be construed as a natural experiment (Diamond 1986). He compared two otherwise similar areas, noted some current differences (grama, mesas, watersheds, etc.), and traced back to known past differences (e.g., overgrazing) that could be reasonably construed as causes of the current differences. With respect to the abundance of deer and other prey, Leopold compared not only those two areas but also other areas in the United States where predators were extirpated during his lifetime, showing the common pattern of boom-and-bust prey cycles in each case where predators were removed (Leopold 1943).

These natural experiments thus allowed Leopold to identify two causes of land sickness in the southwestern United States: overgrazing and predator removal. As he subsequently elaborated, "Overgrazing first mars the plants and then the soil. Rifle, trap, and poison next deplete the larger birds and mammals" (Leopold 1949, 149). But the problem isn't the practices of grazing and hunting themselves; the problem, Leopold suggested, is the way they were done, for how long they were done, and most importantly, *where* they were done. It can take decades rather than years to recover from the damage to a browse range caused by an irruptive deer population in an arid climate like that of the southwestern United States, and the range may wash away before it can recover. As for grazing, Leopold wonders "whether semi-arid mountains can be grazed at all without ultimate deterioration," saying that he knows "of no arid region which has ever survived grazing through long periods of time" (1937, 397). "The trouble is," Leopold stated, "that where water is unevenly distributed and feed varies in quality, grazing usually means overgrazing" (397). Perhaps not surprisingly, the sensitivity of arid regions to human practices likewise became a theme throughout Leopold's writings.

In response to these experiences and others,[1] *land health* came to play an essential role in Leopold's land ethic: "A land ethic, then, reflects the existence of an ecological conscience, and this in turn reflects a conviction of individual responsibility for the health of the land. Health is the capacity of the land for self-renewal. Con-

---

1. In addition to being influenced by his experiences in the Southwest, Leopold was very struck by his experiences in Germany, which he visited in 1935, although he wrote about Europe before the trip. Even though he often used northwestern Europe as an example of disturbed yet healthy land, he noted that foresters in Germany had learned the hard way that multispecies forests were healthier (e.g., in terms of soil) than single-species (e.g., spruce-only) ones. And the deer in this heavily managed, predator-free area were not healthy, suffering from widespread extirpation of their plant browse. See Flader 1994 and Meine 2010 for discussion.

servation is our effort to understand and preserve this capacity" (Leopold 1949, 221). He cited the Southwest, along with the Ozarks and part of the South, as being the locations in the United States that have the worst "disorganization" or "wastage," which he said is "similar to disease in an animal, except that it never culminates in complete disorganization or death" (1949, 219). The reference to the Ozarks is likely a reference to the Dust Bowl of the 1930s, where the removal of deep-rooted prairie grasses in favor of agricultural crops led to massive dust storms when drought set in, rendering the land virtually useless for farming.

But what does it mean for land to have a "capacity for self-renewal"? What causal factors promote (or inhibit) such a capacity, and how do overgrazing and predator removal embody the inhibitory factors? How do we know when land is sick? What are the symptoms? How, if at all, do Leopold's views relate to contemporary concerns over biodiversity and stability and the controversy over the connection between them? Does Leopold offer us a different path out of this puzzle?

Leopold was in the process of writing a book on land health prior to his untimely death (Warren 2013), so unfortunately we can only piece together his thoughts up until that point. My interpretation of land health may thus be somewhat less certain than my interpretations of Leopold's concepts in other chapters, where his thoughts were more developed and polished. Warren's approach tracks the progression of Leopold's ideas about land health over time, and this chapter is indebted to her work.[2] I take a more conceptual approach that focuses on key elements of his views about land health, of necessity drawing from different essays (some unpublished), and focusing primarily on those written later in his life.

Here it is important to note that Leopold often used the word *stability* interchangeably with *land health*. For example, he stated that "technologies are preventatives, not cures, and that, applied in time, they will successfully preserve for land its normal stability of organization, or health" (Leopold 1942b, 202). Similarly, he stated, "The Wisconsin land was stable, i.e., it retained its health, for a long period before 1840" (Leopold 1944a, 311).

2. See also Mitman 2005, for broader historical context.

Philosophers have given great attention to the word *stability* in the essay "The Land Ethic," seeing it as part of the so-called summary moral maxim along with *integrity* and *beauty*, and attributing to Leopold views of stability that were held by other ecologists of his time (see chapter 1 for discussion of the "summary moral maxim" and stability). Much of that attention has been critical. But Leopold explicitly rejected the phrase used by some ecologists, *balance of nature*, stating that the phrase "fails to describe accurately what little we know about the land mechanism" (Leopold 1949, 214), so that cannot have been what he meant by stability.

Moreover, Leopold acknowledged that the land was always changing, even changing in response to human activity (as mentioned in the previous chapter), but as the Southwest illustrated, some of those changes were problematic. On the other hand, as noted in the quotation above, Leopold characterized the land ethic in terms of land health. Even though in "The Land Ethic" he doesn't directly state that stability and land health are equivalent, he spends much of the essay explaining the mechanism underlying the land's ability to self-renew as well as the sort of actions that undermine that ability. (And in none of it does he talk about "dynamic equilibrium" or other concepts more commonly associated with ecological stability).

Thus, a plausible interpretation is that stability is land health (or something very close to it).[3] In what follows, I proceed under the assumption that land health and stability can be used interchangeably in Leopold's writings. With this assumption, to the extent that the so-called summary moral maxim (which references stability) expresses desired outcomes for land communities, the outcome to be sought is land health. Furthermore, as noted previously, Leopold explicitly stated that "[c]onservation is our effort to understand and preserve this capacity" of the land for self-renewal (1949, 221).

---

3. Warren offers a slightly different interpretation: "In Leopold's lexicon, in short, stability ranked as a fundamental characteristic of healthy land. Indeed, so linked was the land's health to its stability—to its sustained ability to recycle nutrients and maintain soil fertility—that Leopold several times treated the words 'health' and 'stability' as if they were synonyms.... The words were not synonyms, though: 'stability' had a more narrow, particular meaning, whereas 'health' was a broader concept, not yet well grasped. As Leopold understood land health, however, stability was a defining element; if stable land was not fully healthy, it was very close" (Warren 2013, 427).

In short, understanding land health is crucial for understanding Leopold's land ethic.

Here one might wonder why I focus on stability rather than integrity, given that the latter also appears in the so-called summary moral maxim (which again, I am de-emphasizing).[4] Aside from the more extensive discussion of stability in the context of land health across several essays (as discussed throughout this chapter), I find that the term *integrity* is the more straightforward term and thus less in need of analysis. In "The Land Ethic," Leopold referenced integrity in the context of stating that all species, economically valued or not, are "entitled to continuance," noting that the "stability [of the land community] depends on its integrity" (1949, 210). Similarly, in "The Land-Health Concept and Conservation" Leopold elaborated on an instruction to "cease throwing away [a land community's] parts" with a section entitled the "Integrity of the Parts," once again discussing integrity in the context of the membership of land communities: that is, what species they contain (Leopold 1944b, 220–21). These references suggest that by *integrity*, Leopold meant simply the retention of species within a land community, and as is discussed below, in his view, this was indeed important for land health.[5]

I begin with an exploration of what Leopold meant by land health and what he saw as its symptoms, causes, and underlying mechanism. I then compare Leopold's approach to contemporary debates over the connection between biodiversity and stability, address potential concerns regarding Leopold's conception of land health, discuss its philosophical and scientific significance, and conclude.

UNCOVERING LEOPOLD'S VIEWS ABOUT LAND HEALTH/SICKNESS

In the "Wilderness" essay in ASCA, Leopold asserted that the most important characteristic of an organism is its *health*: the capacity for

---

4. For elaborations on Leopold's land ethic that do focus on integrity, see Norton 1992 and Westra 2001.

5. Warren (2013) offers a similar characterization, noting that when using the term, Leopold typically wanted to emphasize the species that were important for stability/land health while acknowledging that we might be ignorant of which ones those were.

internal self-renewal (Leopold 1949). He further thought that two organisms whose processes of self-renewal had been subjected to human interference and control were humans and land (Leopold 1949), where land includes soils, water systems, "wild and tame" plants and animals (Leopold 1942a, 199).[6] With respect to the land in particular, Leopold saw health as a state of vigorous self-renewal in each component of the land and in all collectively (Leopold 1944a). He characterized this self-renewal as the "collective functioning of interdependent parts for the maintenance of the whole" (1944a, 310). This collective functioning, Leopold goes on to suggest, is characteristic of an organism; moreover, "[i]n this sense land is an organism, and conservation deals with its functional integrity, or health" (310).[7]

Symptoms of Land Sickness

Our human experience of sickness usually begins with symptoms; typically, Leopold began discussions of land sickness by listing them. For example, in an essay published in 1946, Leopold listed the following symptoms of land sickness or "disorganization":

1. Abnormal erosion (soil washing or blowing away)
2. Abnormal intensity of floods
3. Decline of yields in crops and forests (perhaps as a result of loss of soil fertility, as he discusses elsewhere) (e.g., Leopold 1944a)
4. Decline of carrying capacity in pastures and ranges
5. Outbreak of some species as pests: for example, deer irruptions (Leopold 1944a)
6. Disappearance of some species without visible cause
7. A general tendency toward the shortening of species lists and food chains
8. A worldwide dominance of plant and animal weeds

These can be understood as *indicators* of land sickness. Note that items 6 and 7 describe the loss of biodiversity. As we will see mo-

---

6. Leopold thus seems to be using the term *health* as applied to land literally, not as a metaphor. More on this below. See chapter 3 for a discussion of land as a biological whole.

7. Here, I see "functional integrity" (the functions remain intact) as different from, although of course related to, "integrity" simpliciter (the species remain intact). *Some* species can be lost and yet functions remain intact.

mentarily, Leopold also thought of loss of biodiversity as a *cause* of land sickness. Presumably, he thought it was important to have a list of symptoms for the same reason that doctors do: in order to identify the presence of illness even when the illness itself is not directly detectable.

A small but important aside: It is generally agreed that the term *biodiversity* was coined around 1985, perhaps by W. G. Rosen, but Leopold's "diversity of flora and fauna" (Leopold 1939a, 730; 1942a, 203) is surely of a similar meaning, especially given the multitude of meanings ascribed to *biodiversity* (see Meine 2004, who traces Leopold's influence on later understandings of biodiversity). I use this anachronism throughout this chapter as an easy shorthand for the contemporary reader, but as with the term *stability*, we need to recognize that Leopold develops his own distinctive approach. In particular, he is not referring to the mere number of species in an area, as will become clear below.

To recognize these symptoms, Leopold maintained that the science of land health needs "a base datum of normality, a picture of how healthy land maintains itself as an organism." (Leopold 1949, 196). This base datum, he thought, could come from places of intensive human occupation, such as northeastern Europe, but more often they were relatively less modified "wilderness" (again recalling that for Leopold, wilderness did *not* imply "free of humans"). Thus, as he did when comparing the mountainous areas of the southwestern United States to Chihuahua, Mexico, Leopold would use a similar "wild" area for comparison to a less wild one. Alternatively, Leopold maintained, one could compare the present to the past via paleontology; he believed that the paleontological record showed that in the past, component species were rarely lost and rarely got "out of hand," and that weather and water built soil as fast or faster than it was carried away (Leopold 1949). (The paleontological record available to us today is much more developed and shows many more complexities).

Causes of Land Sickness

But what are some of the *causes* of the symptoms of land sickness? Leopold thought that we knew with reasonable certainty that probable "maladjustments" of the land community coincided with periods of "violent change" in the land community (Leopold 1944a, 315). He was also fairly certain that a *loss of soil fertility* (due, e.g., to inappropriate agricultural practices) was a cause of land sickness (Leopold 1944a; 1946b; 1949); as previously mentioned, the southwestern United States and the Dust Bowl of the 1930s were prime examples.[8]

However, aside from violent changes and loss of soil fertility, Leopold thought that causes of land sickness were hard to determine. Asserting a causal relation would be to imply that we understood the underlying mechanism, but he believed that the land mechanism is too complex to be understood, forcing us to "make the best guess we can from circumstantial evidence" (Leopold 1944a, 315). The circumstantial evidence available to him indicated an association of twenty thousand years between "stability and diversity in the native community" in Wisconsin, an association that suggested that stability and diversity depend on each other (315). But, "[b]oth are now partly lost, presumably because the original community has been partly lost and greatly altered. Presumably the greater the losses and alterations, the greater the risk of impairments and disorganizations" (315).[9]

In other words, Leopold was saying that we have a strong reason to think that there is a causal relation between soil fertility and stability/land health and a less strong, but still significant reason to think there is a causal relation between biodiversity and stability/land health. Both causes, he thought, exhibited a greater effect the greater the amount of change, and both causes (and corresponding effects) were in play in the southwestern United States.

At this point a worry about circularity might arise, since Leopold has named biodiversity as a cause of stability and loss of biodiversity as a symptom of instability; he might seem to be saying that

---

8. See Flader 2011 for further discussion of Leopold's views on the importance of sustaining soils. Note that the importance of soil fertility to the health of land communities should not be seen to preclude land communities that are water-based. On this point, Leopold wrote, "Soil health and water health are not two problems, but one" (Leopold 1941b, 22).

9. Similarly, he wrote, "What, in the evolutionary history of this flowering earth, is most closely associated with stability? The answer, to my mind, is clear: diversity of fauna and flora" (Leopold 1942a, 203).

biodiverse land communities will be more biodiverse and that less biodiverse land communities will be less biodiverse, which would be circular and thus uninformative. However, I do not think he is being circular. Leopold argued that the loss of some species may lead to the loss of other species because the latter are dependent on the former, either directly or indirectly via the soil. As noted earlier, Leopold was particularly concerned with communities that had lost predators, having witnessed the downstream consequences on deer and various plant species after wolves and mountain lions were extirpated. In other words, the loss of the predator function in particular could lead to loss of land health, predator-prey interactions being the basis for one of the most important of the interdependencies (but not the only one) that merit preservation. A feedback loop between diversity and stability/land health is another possibility; Leopold wonders whether stability and diversity are interdependent (1939a).

But how strong is the causal relation between land health and biodiversity? Can you have one without the other? Leopold believed this could sometimes happen. For example, he thought that northwestern Europe was an intensively used landscape which seemed to have remained stable despite the loss of fauna and flora diversity; its farm soils remained fertile; its water systems still produced many fish, few floods, and little silt; and there were some pest irruptions, but fewer than in the United States. He acknowledged that not all signs were positive: he said, for example, that "migratory game birds are in a bad way" (Leopold 1942a, 204). So, loss of biodiversity does not inevitably imply a loss of land health.

The question is, how common were counter-examples like northwestern Europe? Did "other parts of the globe remain stable without the deliberate retention of diversity"? (Leopold 1942a, 204). Leopold doubted that they did. He thought that other parts of the globe were either "undeveloped (the tropics, the arctics), in process of dislocation (most of United States, South Africa, Australia, China), or already relapsed into a retrograded stability (Mediterranean countries)" (Leopold 1942a, 204).[10] A few years later, he was a

---

10. Leopold does not elaborate here on the phrase "retrograded stability," but it seems likely that he is thinking of land that has a reduced carrying capacity, as discussed elsewhere in this chapter.

little more optimistic. Europe was still the most positive case where stability persisted in the face of biodiversity loss, along with Japan, but "[m]ost other civilized regions, and some as yet barely touched by civilization, display various stages of disorganization, varying from initial symptoms to advanced wastage" (Leopold 1949, 219). In short, Leopold was hypothesizing a causal connection between biodiversity and stability/land health, while recognizing that the connection is defeasible, depending on local circumstances (probably including climate and other factors generally unknown).

The Capacity for Self-Renewal

But what is land health—what is the land's capacity for self-renewal? As early as 1923, Leopold spoke of land's self-healing power—or the lack thereof—after being injured (Warren 2013). In "Land Pathology," he described this insight as a product of a "series of observational deductions" (1935b, 213). Before the machine age, land could "right itself" through "automatic adjustments" such as "population cycles, emigration, starvation, interpredation" (213).[11] Note that these are relatively short-term ecological, not evolutionary adjustments. Since the machine age, many land communities had been unable to make such adjustments. Europe was unusual in its resistance to "abuse," preserving an ability to restore "new and stable equilibria between soil, plants, and animals." The United States, on the other hand, had seen an "accelerating velocity of destructive interactions," but in the United States, in contrast to Europe, "[r]ecuperative mechanisms either do not exist, or have not had time to get under way" (214). To contemporary ears, this all might sound akin to recent discussions about *resilience*; more on this below.

By the time of ASCA, Leopold had honed his account of how self-renewal in land communities operates, using three interrelated ecological metaphors: *land pyramid, food chains, and fountain of energy*.[12] His discussion of them takes up a substantial portion of "The Land

---

11. I am not sure what interpredation is; perhaps it is competition between predators.

12. See chapters 2 and 3 for additional discussion of the three metaphors; see Warren 2013 for a discussion of Elton's influence on these Leopoldian metaphors.

Ethic," drawing from his earlier (1939a) essay "A Biotic View of Land." I explain each of these metaphors in turn.

The *land pyramid* metaphor describes land communities in terms of a pyramid consisting of layers (see fig. 2.2). Soil is the lowest layer, then plants, then insects, then birds and rodents, through various groups (humans are in these middle layers), and on to the apex layer consisting of the largest carnivores. Leopold explained that "[e]ach successive layer depends on those below it for food and often for other services, and each in turn furnishes food and services to those above" (Leopold 1949, 215). The pyramid shape is meant to suggest the relative prevalence at its layer (e.g., predators are typically considerably outnumbered by their prey).

Food chains, on the other hand, are "lines of dependency" for food and other services: for example, rock → soil → alfalfa → cow → farmer → grocer. Leopold thought that each species, including humans, is a link in many chains, so that the pyramid consists of a "tangle" of chains; it seems disorderly, but stable systems imply that it is a "highly organized structure." Leopold thought that the functioning of the system depended on the cooperation and competition of its diverse parts—the performance of what we might call the "role functions" of different species, such as predator-prey interactions and parasite-host interactions (see Millstein 2020b for discussion). Over time, Leopold thought, evolution had added layers to the pyramid and links to the food chains, with "the trend of evolution" in the direction of elaborating and diversifying the biota (1949, 216).

The *fountain of energy* metaphor was characterized in terms of food chains that conduct energy up the pyramid's layers, with death and decay returning energy to the soil (see fig. 2.2). Leopold acknowledged that the energy circuit is not closed, that "some energy is dissipated in decay, some is added by absorption from the air, some is stored in soils, peats, and long-lived forests" (1949, 216). However, he maintained that in a healthy, or stable, land community, "it is a sustained circuit, like a slowly augmented revolving fund of life."

My elaboration of self-renewal/land health here follows Leopold (1949), except where noted.

Such energy circulation depends on the complex structure of the plant and animal community—the characteristic numbers as well as the characteristic kinds and functions (again, interactions) of the component species.[13]

Given the potential for energy loss, Leopold hypothesized that *long food chains* are necessary for continued energy circulation. When food chains are longer, Leopold proposed, nutrients spend relatively more time bound up in organisms—that is, passed from organism to organism through feeding—and less time in the soil and water. When food chains are shorter, on the other hand, nutrients spend less time bound up in organisms, relatively speaking, and more time in the soil and in water. Thus, with shorter food chains, there is more opportunity for nutrients to be eliminated via wind or water erosion of soil, with the soil's nutrients ending up in the ocean rather than recycling (Leopold 1941b; see also Warren 2013). Longer food chains are better able to retain nutrients within the land community. Yet human tools had enabled shortening of food chains, most obviously with the removal of predators but also through other means such as the polluting or damming of waters, which sometimes led to extinction of plants or animals. These violent, rapid, and large-scale changes threatened land health.

Bringing these ideas together, Leopold thought that a many-layered pyramid and a tangled web of long food chains, representing diverse species populations interacting in ways that transmit matter and energy, would result in a sustained fountain of energy (continued energy circulation) that enabled the self-renewal capacity that Leopold dubbed land health. When there is "continuity of this organized circulatory system," when "its food chains are so organized as to be able to circulate the same food an indefinite number of times," the land is able to self-renew (Leopold 1942a, 205). There is "collective functioning of interdependent parts for the maintenance of the whole" (1944a, 310). The structure "seems to function and to persist" (1939a, 729), preserving its habitability for humans and most other species. The land can support a diversity of life over time, season after season, as in the Sierra Madre. In this sense, Leopold's concept

---

13. This, then, is the sense of biodiversity that Leopold thinks is necessary for land health, and not just the mere number of species. More on this below.

of land health presages some aspects of our contemporary notion of sustainability; he even speaks of the energy circuit as "sustained" (1949, 216) and applies the same term to self-renewal (1942c, 487).

A PROPOSAL FOR UNDERSTANDING LEOPOLDIAN LAND HEALTH

To summarize, I propose that Leopold thought *land health* was

> a land community's capacity for self-renewal, or stability, which depends on the ability of energy to continue to cycle within the land pyramid, which in turn depends on *biodiversity* (retaining species in their characteristic numbers, kinds, and functions/interactions, hypothesized to form long food chains facilitating the continuous circulation of food and nutrients) and on *soil fertility*, resulting in the ability of the land community to support a diversity of life over time.

Remember, however, that Leopold thought the evidence for soil fertility as a cause of land health was stronger than it was for biodiversity as a cause of land health, and he acknowledged known exceptions (e.g., Western Europe) to the former.

Land sickness, according to Leopold, "never results in complete disorganization or death"; the land can recover, but typically it would do so at some "reduced level of complexity, and with a reduced carrying capacity for people, plants, and animals" (1949, 219), as in the southwestern United States. This unstable land, he said, "can no longer recirculate the same food an indefinite number of times," and he viewed such symptoms as "[e]rosion, floods, pests, loss of species, and other land-troubles without visible causes" as "expressions of this instability" (Leopold 1942a, 206). He thought, although he admitted he could not prove or disprove it, that "deliberate retention of both fertility and diversity" would reduce instability (204).

Thus, Leopold concluded that the changes effected by humans should be less drastic, of smaller scope, and less rapid in order to retain land health/stability, especially since we generally cannot predict the outcomes of our actions. Land communities need to retain soil fertility and a diversity of flora and fauna (Leopold 1942a) to persist, to continue to support life over time. In this vein, Leopold famously remarked, "To keep every cog and wheel is the first precaution of intelligent tinkering" (1938b, 416).[14] Here and elsewhere, we see Leopold expressing the importance of *integrity* (which, like stability, is included in the "summary moral maxim"), since, "[a]s far as we know, the state of [land] health depends on the retention in each part of the full gamut of species and materials comprising [the land community's] evolutionary equipment" (1942b, 300). More succinctly, Leopold stated that the land community's "stability depends on its integrity" (1949); evidence suggested that the integrity of native communities and their ability for self-renewal have "some causal connection" (1946b, 514).

Having said above that it is reasonable to understand Leopold's idea of land health as a form of sustainability, one might also wonder if it can be understood as *resilience*. In recent years, there has been an explosion of papers exploring different definitions and measurements of resilience in ecology, with Holling (1973) usually cited as the inspiration; Chambers et al. (2019) provide one recent summary. It is an interpretive challenge, but on the whole I think it would be a mistake to understand Leopold's land health as resilience. Although resilience concepts in the literature vary widely, what they all seem to have in common, following Holling (1973), is the assumption that the system is responding to a disturbance. So, for example, Berkes et al. (2012) characterize resilience as "the capacity of a system to *absorb disturbance* and reorganize while undergoing change so as to still retain essentially the same function, structure, identity, and feedbacks" (280; emphasis added). They make the further claim that land health can be "reinterpreted and extended through a resilience lens" (280). Yet, as discussed above, Leopold thought that land could be healthy even if it could not withstand "violent" or "rapid" changes (disturbances); presumably, the healthy Sierra Madre could not have withstood them any better than the southwestern United States did, given that they both shared the delicate constitution of an arid climate. Western Europe, on the other hand, was both healthy and resilient. Along these lines, in "Biotic Land Use" Leopold suggested, "The question in hand is whether other parts of the

---

14. Some see this as an early expression of the precautionary principle (e.g., see Newman et al. 2017).

globe [besides Europe] can *remain stable* [i.e., healthy] without the deliberate retention of diversity. All I can say is that I doubt it. Land is unequally sensitive" (1942a, 204; emphasis added).

Thus, he concluded that "[l]ands differ in their toughness," varying in the length of time they remain healthy in the face of violent changes (1942a, 207). So, my claim is that, for Leopold, land health and resilience were separate ideas,[15] although clearly he was thinking about both issues, using "toughness" or "sensitivity" to express "resilience." He thought that in some areas, land was healthy *and* resilient (able to withstand disturbance), while in other areas, it was healthy but *not* resilient (not able, or less able, to withstand disturbance), just as we might say of a person who is currently healthy that they get sick easily, while another healthy person hardly gets sick at all. The difference is their response to disturbance (e.g., exposure to certain viruses or bacteria). The two ideas are thus distinct—although obviously related—in his thinking. And I think there is a benefit to keeping them distinct; separating land health from resilience has the advantage of identifying certain regions (such as arid ones) as in need of especially gentle treatment if they are not resilient, even if currently healthy.

Here as elsewhere, though, it is important to reiterate that Leopold would be humble about our ability to determine which land communities are resilient and which are not, necessitating careful treatment in order to avoid damaging land health. The fact that a land community is currently healthy does not imply that we can disturb it without consequence, because that healthy land community might not turn out to be resilient.

CONNECTING TO CONTEMPORARY BIODIVERSITY-STABILITY DEBATES

The topics of this chapter, like others discussed in this book, might cause some to worry that Leopold's views of biodiversity, stability, and the connection between them are idiosyncratic or have been shown by contemporary science to be false. The brief discussion

---

15. If I am correct in my interpretation, this differentiates Leopold's view of land health from the view of ecosystem health view held, for example, by Costanza and Mageau 1999.

in this section is meant to allay such worries. If anything, Leopold's views on these issues seem quite prescient; some scholars have only recently accepted them after finding problems with other approaches. As a consequence, Leopold's views on biodiversity, stability, and their connection ought to be taken quite seriously by contemporary scholars.

Numerous papers have contributed to the biodiversity-stability debate—far more than I can summarize here. Thus, I rely on the excellent historical account of deLaplante and Picasso (2011), to which I refer the reader for further detail. DeLaplante and Picasso identify three historical periods of this shifting debate: (1) 1950s and 1960s— The view that diversity is positively correlated with stability was endorsed by a number of prominent ecologists, including Odum, MacArthur, and Elton. (2) 1970s and 1980s—The consensus shifts away from this view toward thinking that diversity is *not* correlated with stability, as in the work of prominent ecologists May and Pimm. (3) 1990s—Led by David Tilman and others, largely focusing on plant experiments, the diversity-stability hypothesis transforms toward seeing a positive relationship between diversity and the stability of various functionally defined properties of communities and ecosystems.[16] (These functionally defined properties include resistance to invasion by new species and the temporal stability of an ecosystem property like biomass productivity). Importantly, throughout this period different conceptions of "biodiversity" and "stability" are in play, with the result that ecologists who appear to disagree with one another may in fact not be doing so directly. Similarly, some concepts are quite different from Leopold's: for example, biomass productivity is quite different from Leopold's land health concept and may not correlate with it at all.

Then at the turn of the twentieth century, the debate flared up, only to cool down again, beginning in 1999 with a panel of ecologists reporting in the *Ecological Society of America Bulletin* that there was scientific evidence that loss of biodiversity impacted ecosystem functioning by reducing plant productivity, which de-

---

16. Recall from chapter 3 that Leopold's land community concept is not fully synonymous with the ecosystem concept, but it is close. It is a blended concept of ecosystem and ecological community that incorporates both matter and energy flows as well as species interactions.

creased ecosystem resistance. In mid-2000, a group of critics of the biodiversity-ecosystem function experiments wrote a letter to the *ESA Bulletin* heavily criticizing the report, accusing it of bias and propaganda. But in late 2000, a synthesis conference was held in Paris to try to reconcile different interpretations of empirical results. This conference was widely viewed as successful and as having calmed some of the vitriol.

Some outcomes of the synthesis conference, according to deLaplante and Picasso (2011), include general agreement on the following points:

- A large number of species ("species richness") is required to maintain ecosystem functioning, but whether this is because more rich communities have some key species that differentially affect ecosystem function, or because diversity effects arising from niche complementarity had an effect on ecosystem function, was unclear.
- It is unresolved whether most extinctions are random, but the answer affects the realism of some models and experiments.
- A greater number of species may be needed to maintain stability in ecosystems (the "insurance hypothesis").
- Observational and experimental data on diversity-productivity relationships differ and need to be reconciled.
- There are limitations to the generalizations about other ecosystems (e.g., aquatic) and other trophic levels (e.g., consumers, decomposers) that we can draw from experimental evidence about grasslands ecosystems.
- The functional traits of species and their interactions are the factors that predominately affect ecosystem functioning.

These points of general agreement provide goals for further studies and some "second generation" biodiversity experiments have indeed followed up on these issues. For our purposes, what is particularly notable is how many of these points dovetail with Leopold's views: for example, the need to consider a full range of trophic levels and to look for realistic "models" for conclusions more generally; a recognition of the potential importance of an "insurance hypothesis" (a precautionary approach will seek to save "every cog and wheel"); the importance of examining species functions and interactions with consideration that some species' roles may be especially important. In other words, in some respects, Leopold was already going in a direction that many—but admittedly not all—ecologists today think the field should take.[17]

What is this alternate direction? Table 4.1 shows some broad differences in approach between Leopold and many contemporary ecologists. Some contemporary ecologists are now arguing for a return to Leopold's approach with respect to methodology, species studied, and the other respects outlined in the table.

One question that arises from this table is whether in studies of potential diversity-stability connections, the pendulum has swung too far toward field experiments and theoretical models in an understandable attempt to introduce rigor to the question. The rigor and precision of mathematical models and field experiments—especially with a limited range of species—may be failing to take into account the very dynamics that matter for a diversity-stability connection. Leopold's historical, observational, and comparative approach (including "natural experiments"), some of which involved hands-on knowledge about what works and what doesn't (e.g., eliminating predators, grazing in arid climates), could provide additional perspective to the debate, especially when coupled with his richer notions of biodiversity and stability.

Moreover, some contemporary scholars are embracing what appear to be some very Leopoldian approaches—notably, a focus on the loss of biodiverse species interactions (Valiente-Banuet et al. 2015), an emphasis on "processes or attributes that contribute to the self-maintenance of the ecosystem, including energy flow, nutrient cycling, filtering, buffering of contaminants, and regulation of populations" (Thompson et al. 2012, 689), and the study of natural systems consisting of many trophic levels (Andresen et al. 2018). In addition, many scholars are examining systems that include humans under the rubric of social-ecological systems (e.g., Ostrom 2009; Berkes et al. 2012). These studies provide additional support for Leopold's ideas, showing that they are worthy of further pursuit.

For example, what of Leopold's hypothesis concerning long food chains in a complex web of interdependencies as an underlying

---

17. Although today's ecologists who go in Leopold's direction do so for perhaps not entirely independent reasons, given the influence on Leopold from Elton and Leopold's influence on Odum and the field more generally, and given that Odum and Elton have both been extremely influential overall.

TABLE 4.1. Comparison of Leopold to contemporary ecologists with respect to topics related to biodiversity and stability

Biodiversity/ stability issue	Most contemporary ecologists	Leopold
Methodology	Mostly field experiments or theoretical models	Historical, observational, and comparative approach; natural experiment; hands-on practitioner
Species studied	Many on plants only	All trophic levels
Mechanisms	Not always studied; multiple mechanisms proposed	May never be fully understood; proposes long food chains in complex web of interdependencies in a many-layered land pyramid as the basis of sustained matter/energy flow
Biodiversity	Most commonly use species richness, at least as a proxy	Species collectively manifest a diversity of interactions/ interdependencies within food chains
Stability, broad meaning	Succeeded by ecosystem function	Capacity for self-renewal, ability to support a diversity of life over time
Stability, manifested/ measured	Often measured by productivity—the rate of generation of biomass in an ecosystem	Arises from land community functioning—the performance of interactions between species populations
Self-maintenance	Often excludes humans	Includes humans

mechanism for land health? Hooper et al. state that "identifying mechanisms of biodiversity effects" is an area of uncertainty needing further research, and that resolving "relationships among taxonomic diversity, functional diversity, and community structure" is important for doing so; moreover, they explicitly argue for long-term field research that examines multiple trophic levels (Hooper et al. 2005, 4). Are long food chains worthy of further exploration? Long et al. (2011) note that a number of studies have examined connections between longer food chains and stability, and that these have had conflicting results. However, many of these studies model, either theoretically or experimentally, additions of species to ecosystems. Long and colleagues do this, examining the effects of adding predators to a marine intertidal food web. Yet as noted above, it was Leopold's view that food chains were largely formed by evolution, with ecological adjustments in response to changes varying in their success. Indeed, harmful invasive species were already known to Leopold; see, for example, his "Cheat Takes Over" essay in ASCA. So, adding species to land communities might not be the best test of his version of the hypothesis; rather, further observational and comparative studies, especially of systems that have undergone known perturbations, could be a more fruitful and realistic test.

Thus, far from being outdated, Leopold's views on land health offer a promising conceptual apparatus for future work and environmental policy, especially given the continuing influence of his land ethic. That being said, it is important to recall that these are just hypotheses for Leopold, based on what he knew. He might not have been opposed to considering other forms of biodiversity-stability or other possible mechanisms. But he was worried about human changes that affect the ability of land communities to maintain a diversity of life over time, a core concern for environmental ethics and conservation biology. And lest it get lost among more controversial hypotheses about biodiversity and stability, the concerns that Leopold raised about the loss of soil fertility are all too timely in the present day.

ADDRESSING POTENTIAL CONCERNS WITH LEOPOLD'S LAND HEALTH CONCEPT

In the previous section I suggested that there are reasons to think that Leopold's conception of land health is still very much a live option for contemporary ecology and conservation biology. Nonetheless, some concerns have been raised about other understandings of ecosystem health, and understandings of health more generally, and so it is worth discussing if they apply to Leopold's land health conception as well (even if land community and ecosystem are not precise synonyms, as discussed in chapter 3).[18] In this section, I argue that they do not.

One concern is that ecosystems are not organisms and so cannot be literally healthy (see, e.g., McShane 2004 for discussion). As noted above, Leopold sometimes did refer to the land as an organism and stated that it manifested health in the same way: the capacity for internal self-renewal. Although his usage was not consistent, to the extent that he did think of land communities as organisms, they can be seen as being literally, and not just metaphorically, healthy or not. On that interpretation, was that a reasonable position for him to have held? I think a case can certainly be made that it is a reasonable position. The concept of *organism* has recently undergone a fair amount of philosophical reexamination, given challenging cases such as holobionts (usually conceived of as hosts together with their resident microorganisms), and some of the resulting conceptions might lend themselves very nicely to being applied to land communities/ecosystems. For example, Subrena Smith (2017) argues that organisms should be understood as "functionally differentiated and integrated persisters," a characterization that seems in accordance with Leopold's land community concept. This is a less restrictive conception of an organism than is usually assumed when concerns are raised about seeing ecosystems as organisms; for example, it doesn't assume that there are regimental developmental stages, a "superorganism" idea associated with Clements (but see Eliot 2011). Smith herself suggests that organisms that "are nodes in a complex web of dependencies" (2017, 12) might form larger-scale organisms that are organisms to a lesser degree (given that they are less stable and less integrated than the organisms that compose them), but that should still mean that such organisms, in the form of land communities, can be bearers of

---

18. See Dussault 2021 for a more extensive discussion of the problems underlying some of these objections; my account here is broadly sympathetic to Dussault's.

health (entities to which the concept of health can be legitimately applied).

However, even if ecosystems or land communities are not organisms, an argument can be made that there are legitimate conceptions of health that are applicable to entities that are not organisms (e.g., McShane 2004 and Dussault 2021 argue this). Indeed, Leopold's understanding of land health shares many aspects of other accounts of health more generally. Like Boorse (1977) and Wakefield (1992), Leopold invokes functions and functioning in his conception of health (see Millstein 2020 for an elaboration and defense of Leopold's use of functions in terms of coevolution and a "selected effects" account of function). Like the World Health Organization's characterization of "health," Leopold saw land health as a positive state and not just the absence of disease. Whitbeck (1978) and McShane (2004) argue that the concept of health is value-laden and normative; Leopold (1944a, 316) speaks of the components of the land having a "collective as well as a separate welfare" that conservation must deal with. Kingma (2010) argues that accounts of health must take into account the ways that bearers of health are dynamic and responsive; as noted above, Leopold described the sorts of adjustments (species can change their numbers or their behavior, etc.) that land communities can make in response to changes to the energy circuit. Indeed, perhaps thinking about health has come full circle, with human health defined as "the ability to adapt and to self manage," inspired by the way that "environmental scientists describe the health of the earth as the capacity of a complex system to maintain a stable environment within a relatively narrow range" (Huber et al. 2011, 3).

Another concern is based on the claim that ecosystems are not entities and lack objectively real boundaries. Some authors think that being an entity with objectively real boundaries is necessary for being a bearer of health, although not all do (see McShane 2004 for discussion). The question of boundaries of land communities is addressed in chapter 3, where I argue for a Leopoldian conception of land communities whose boundaries are formed by differentials in interactions and matter/energy flows. I refer the reader to that chapter for further details. If the argument of that chapter succeeds, then this concern is moot.

Shrader-Frechette (1997) has raised a different sort of concern, arguing that the concept of ecosystem health is vague and thus not useful; we could, she argues, adopt more precise criteria, but then we would no longer need the concept of ecosystem health. Callicott (1997) and Dussault (2021) offer helpful replies, pointing out, for example, that the concept of human health is similar, yet plays an important role by providing a general term that captures all of the more specific aspects that comprise human health (body temperature in a certain range, blood pressure in a certain range, etc.). Along those lines, I think that Leopold's conception of land health as the capacity for self-renewal can be seen as a broad term that can be further specified by various mechanisms that might underlie it. Leopold proposed one—long food chains that can sustain matter and energy flow—but I think it is clear that he was open to others. The broad term allows for the possibility of other mechanisms that might promote the same or similar capacities, and that is a point in its favor. We can test Leopold's own proposed mechanism and various others that might be proposed. Leopold's land health concept is, admittedly, not a quantitative concept. To that extent, it is vague, but Leopold's helpful list of symptoms can serve as indicators for land sickness, much as lists of symptoms do for illness in humans. Moreover, the examples of the southwestern United States and the Dust Bowl show that there are clear cases of actions that lead to the loss of land health, such as overgrazing and killing off predators in arid climates, leaving the land at a "reduced level of complexity, and with a reduced carrying capacity for people, plants, and animals" (Leopold 1949, 219).[19] Leopold's land health concept is suitably informative and useful.

Finally, Lackey (1996) raises the concern that to say that an ecosystem is healthy is just a way of expressing a state that is desired or preferred. As Dussault (2021) points out, this risks making ecosystem health relativistic by, for example putting polluter preferences on par with conservationist ones. Callicott's (1995) response to this concern is to characterize ecosystem health in terms of an

---

19. Indeed, a recent analysis describes the long-term effects of the Dust Bowl as not only ongoing but permanent (Hornbeck 2012).

ecosystem's linked processes and functions occurring normally or changing normally, where *normally* is understood as *historically*. Yet although Leopold seems to have thought that, historically, most land was healthy (according to the historical records that were available to him), his definition of land health is not tied to that. As noted previously, on his view land could change substantially and still be healthy, although typically (with notable exceptions) those changes occurred slowly. So "land health" does not tie land down to any particular historical period; preserving or restoring land health does not require faithfulness to a set of species or species interactions from some point in time. Instead, we can recognize a sense of normativism where judgments about health are based on judgments about the well-being of the entities whose health status is being assessed (a view described but not endorsed by Dussault [2021]); I think Leopold's views fall into this category. Land health is, first and foremost, good for the land community as a whole, allowing it to persist and thrive. It may also be—it is desirable for it to be—good for the members of the community, and in many cases those goods will align, but that is not required. Indeed, a laser focus on increasing a single species, deer, ultimately led to land sickness in the land communities where predators were eliminated (e.g., in Germany), a state of affairs that was bad for the deer in the long run. So, human beings may or may not "prefer" or "desire" healthy land communities; the value-ladenness of the term "land health" lies not in our preferences but in what is good for the community, remembering that Leopold also recognized the rights of individuals and did not see land health as the only goal of ethics more generally (see chapter 1 for discussion).

PHILOSOPHICAL AND SCIENTIFIC SIGNIFICANCE OF LEOPOLD'S LAND HEALTH CONCEPT

The discussion in the last section—the clarification that "land health" characterizes what is good for a land community—sets the foundation for one of the reasons why Leopold's land health concept is philosophically significant. Following an influential paper by Goodpaster (1978), an entity can only be said to be morally considerable—that is, deserving of moral respect, part of the moral sphere—if it has *interests*.[20] According to Goodpaster, an entity has interests if it is capable of being harmed or benefited, of having a good or bad of its own. Although Goodpaster does not defend the view that ecosystems are morally considerable, he at least leaves the door open, acknowledging that "[t]here is some evidence that the biosystem as a whole" exhibits behavior that would satisfy a plausible definition of life: "self-sustaining organization and integration in the face of pressures toward high entropy" (1978, 323).[21] The rest of the essay goes on to make the case that living beings do have interests in the above sense and thus are morally considerable. Setting aside the question whether land communities are living or not (and whether the definition of "life" that Goodpaster endorses is a good one), if we simply use Goodpaster's criteria for life, then it seems clear that an organized land pyramid with its integrated web of food chains that can sustain itself over time fits the bill. Thus, actions that promote the sustainability of land communities—yielding land health—can be said to be good for those communities, whereas actions which hinder their sustainability—yielding land sickness—can be said to be bad for those communities. In this way, a case can be made that land communities are morally considerable, with land health being the central concept that illuminates how.

To be clear, perhaps Leopold isn't making an explicit argument that land communities can be benefited or harmed and are thus morally considerable entities. But I do think it's reasonable to think that such an argument is implicit. Why else describe at great length the various symptoms of land sickness and the various ways it has been manifested? That being said, I don't think that this is Leopold's only argument for why land communities are deserving of moral consideration. The further argument, what I call an argument from consistency, is described in the next chapter.

Land health is also scientifically significant. As we have al-

---

20. Similarly, Paul Taylor uses the criterion of an entity's having "a good of its own which moral agents can intentionally further or damage by their actions" with "good of its own" understood to mean that "it can be benefited or harmed" and "without reference to any *other* entity" (Taylor 1981, 1999). But cf. Dussault 2018.

21. It's also worth noting that Goodpaster begins his 1978 essay with a quotation from Leopold and that he is more explicit with his support in Goodpaster (1979).

ready seen, it is a promising candidate for understanding what a biodiversity-stability connection might mean, and Leopold's proposed mechanism for land health is a promising hypothesis for further exploration. In addition, Leopold's land health concept can be used as a way to think about various land community situations. (I discuss policy and decision-making implications of the land ethic in chapter 6).

First, we might simply consider some relatively clear-cut, contemporary examples of land health and land sickness. The restoration of wolves to Yellowstone National Park in Wyoming in 1995 is often seen as inspired by Leopold,[22] and indeed, one of the first packs to form was dubbed the Leopold Pack. According to Ripple and Beschta, during the seven decades that wolves were absent from Yellowstone (1920s to the mid-1990s), the formation of new woody browse individuals in species such as aspen, willow, and cottonwood "quickly ceased, with concurrent impacts on soils, beaver, and other ecosystem conditions" (Ripple and Beschta 2005, 617). But after "the reintroduction of wolves, top-down trophic cascades have been observed, including altered patterns of ungulate herbivory, declining elk and coyote populations, new recruitment of woody browse species, and increases in the number of active beaver colonies on the northern range" (Ripple and Beschta 2005, 618), with the beavers themselves having "important roles in the hydrogeomorphic processes of decreasing streambank erosion, increasing sediment retention, raising wetland water tables, modifying nutrient cycling, and ultimately influencing plant, vertebrate, and invertebrate diversity and abundance in riparian ecosystems" (Ripple and Beschta 2012, 211). Assuming these findings are correct—and such studies are not without controversy: Fleming's (2019) critique of Beschta et al. (2018) is just one of many examples[23]—it would indicate that

---

22. Indeed, Leopold stated that "[p]robably every reasonable ecologist" thinks that wolves should be in "the larger national parks and wilderness areas; for instance, the Yellowstone and its adjacent national forests," asking explicitly why wolves that were extirpated from Wyoming and Montana were not used to "restock" Yellowstone instead (Leopold 1945, 322). A few years later, he noted that "Yellowstone has lost its wolves and cougars, with the result that elk are ruining the flora, particularly on the winter range" (1949, 196).

23. Much, although not all, of the controversy has to do with the possibility that other causes are in play, a complexity that Leopold himself would have

with the reintroduction of wolves, the land was healthier. That is, it was better able to support nutrient cycling and a greater diversity of species over time with fewer symptoms of land sickness (e.g., less erosion).

Another clear-cut contemporary example also involves the loss of a top predator and a trophic cascade: overfishing of cod and other predators in the Northwest Atlantic led to the collapse of the benthic fish community, which led to an abundance of small pelagic fishes and benthic macroinvertebrates (predominantly northern snow crab and northern shrimp), once among the primary prey of the benthic fish community (Frank et al. 2005). As a consequence, there was a significant reduction of herbivorous zooplankton and a reduction in nitrate concentrations, a major limiting factor in marine systems. Attempts to restore the system have failed, and it is unclear whether the predatory fish can be restored (Frank et al. 2005). Although there are some promising recent signs, climate change is complicating recovery (Pershing et al. 2015). The land community (which includes water-based communities) has a reduced carrying capacity and shows signs of land sickness.

Beyond these obvious sorts of cases, examples could be multiplied many times, so I will just gesture at various types, although each of these raises many complex issues (often dependent on the empirical situation) that I will likewise just gesture at. Lack of soil health, already an obvious problem in Leopold's time and the other key factor in land sickness that Leopold identified in addition to loss of biodiversity, has become a global problem. Jian et al. note that soil health "represents the ability of soils to function as a biodiverse organism that sustains terrestrial life" but that "soil degradation due to natural vegetation removal, intensive agricultural operations, and erosion are among the main factors causing declines in soil health and crop yield" and that "one-third of soils in the world are infertile due to unsustainable land-use management practices" (Jian et al. 2020, 1). So Leopold was, unfortunately, quite prescient in his concern regarding soil health. Various agricultural practices under discussion today that build soil health and biodiversity would likely

---

appreciated, and which he acknowledged in his own discussion of the effects of removing predators on deer populations (Leopold 1943).

have been ones he would have endorsed, such as the use of cover crops, no-till methods, and hedgerows.[24]

Another type of common contemporary case involves *invasive species*, with ecologists debating the meaning of the term and its connection to other terms like *nonnative* and *non-Indigenous*. Leopold was also well aware of the problem of invasive species.[25] He wrote, "All too familiar are those symptoms of land-illness caused by the importation of exotic diseases and pests," citing examples such as chestnut blight, gypsy moth, and the corn borer (Leopold 1944a, 314). Yet he also recognized "the damage done by control operations" as well as "native plants and animals" that "have assumed all the attributes of pests" (314). And again, the case of Western Europe is relevant here, where "many new plants and animals are introduced" but "the new structure seems to function and to persist" (Leopold 1949, 218). Leopold's underlying message is that the important issue isn't what is native or nonnative per se; rather, the issue is land health (although native species might be less likely to be the cause of land sickness without other perturbations, such as the removal of predators). Thus, we should be exceedingly careful that our actions to address invasive species don't cause further land sickness. Moreover, "there is no such thing as good or bad species; a species may get out of hand, but to terminate its membership in the land by human fiat is the last word in anthropomorphic arrogance" (Leopold 1942c, 487). Clearly, Leopold saw that addressing invasive species would be a complex affair.

Of course, few if any environmental challenges loom as large as

---

24. Leopold (1939b, 420) wrote,

Can a farmer afford to devote land to woods, marsh, pond, windbreaks? These are semi-economic land-uses—that is, they have utility but they also yield non-economic benefits.

Can a farmer afford to devote land to fencerows for the birds, to snag-trees for the coons and flying squirrels? Here the utility shrinks to what the chemist calls "a trace."

Can a farmer afford to devote land to fencerows for a patch of ladyslippers, a remnant of prairie, or just scenery? Here the utility shrinks to zero.

Yet conservation is any or all of these things.

25. See Simberloff 2012 for an extended discussion of Leopold's views on this issue, tracking the changes Leopold's thinking underwent over time.

global climate change, a phenomenon that Leopold would not have been aware of. Yet I think it is obvious that climate change causes land sickness in local land communities as well as meta–land communities (see chapter 3) and the biosphere. There are direct causes, such as species who find themselves out of their livable temperature range—not just from the change in average temperatures but also from changes in temperature extremes (Román-Palacios and Wiens 2020)—and can't migrate or evolve fast enough to survive, as well as indirect causes, such as drought, fire, flooding, and ocean acidification. All of these challenge the ability of land communities at all scales to self-renew and support a diversity of life over time. But perhaps the more significant contribution that Leopold's land health concept will have regarding our thinking about climate change will have to do with potential solutions for addressing climate change. Some proposals for eliminating fossil fuel use in favor of renewable energy might themselves have negative impacts on land communities; two examples are wind turbines that kill significant numbers of birds and solar "farms" located on rare habitat for sensitive species. Likewise, reforestation can be a promising piece of the story, but planting trees (or certain tree species) in certain regions may not make ecological sense; monocultures can be especially problematic. A Leopoldian perspective on land health would caution us to be careful about where and how we address climate change, lest we worsen one devastating problem while attempting to solve another (and in truth, the problems are linked, since the planet's ability to draw down carbon is connected to the health of its land communities).

CONCLUSION

Struck by his experiences in the southwestern United States and the Sierra Madre, his experiences in Germany, and his experiences as a forester and wildlife manager, over the course of his life Leopold became concerned that much of the land he had seen was not healthy. Although he was still working on developing a land health concept at the time of his death, he had come to see land health as the capacity for the land's self-renewal. Underlying that capacity, Leopold believed, was the complex web of food chains (made up of species interactions and resulting interdependencies) arranged in a metaphorical land pyramid, with energy and nutrients flowing up the pyramid from the soil and eventually returning to the soil. Leopold hypothesized that, with longer food chains, land communities could persist—sustain biodiverse land communities for longer periods of time—because nutrients would be bound up in organisms rather than in the soil, where they were always at risk of loss through wind or water erosion. Leopold's ideas are consonant with a number of contemporary approaches to debates over biodiversity and stability and suggest paths for further understanding and study. Although various philosophical concerns have been raised about various ecosystem health concepts, Leopold's land health concept avoids them. Moreover, it is both philosophically and scientifically significant: it shows how land communities can be morally considerable and provides a lens through which to assess whether land communities need protection, restoration, or other human interventions.

There is good reason to think that Leopold saw land health as the primary goal of the land ethic, especially once one understands that stability and land health are essentially synonyms for him. But why should we humans be obligated to try to bring about land health? The next chapter addresses this topic.

CHAPTER FIVE

Arguing for the Land Ethic

Do we not already sing our love for and obligation to the land of the free and the home of the brave? Yes, but just what and whom do we love? Certainly not the soil, which we are sending helter-skelter downriver. Certainly not the waters, which we assume have no function except to turn turbines, float barges, and carry off sewage. Certainly not the plants, of which we exterminate whole communities without batting an eye. Certainly not the animals, of which we have already extirpated many of the largest and most beautiful species. A land ethic of course cannot prevent the alteration, management, and use of these "resources," but it does affirm their right to continued existence, and, at least in spots, their continued existence in a natural state.

ALDO LEOPOLD, "The Land Ethic"

INTRODUCTION: ALDO LEOPOLD'S COMMUNITIES

On April 14, 1948, Aldo went into his office to catch up on correspondence with friends and colleagues.[1] He received a phone call from Oxford and learned that the book that would become *A Sand County Almanac* (hereafter ASCA; he was calling it *Great Possessions* at the time) would be published, some very welcome news that he shared with his wife Estella and one of his daughters, Estella Jr. Subsequent days brought more correspondence, including with his book's illustrator-to-be Charlie Schwartz and his son Luna.

Then Aldo took a trip out to the Shack (a rebuilt chicken coop where the Leopolds spent many weekends) in rural Sauk County, Wisconsin, with the two Estellas, planning on beginning their spring tree planting. They also took time to appreciate the local wildlife, including glorious displays of hundreds of geese.

On the morning of April 21, Aldo was with his family members at the Shack, repairing some tools, when they saw smoke coming from the house of a neighboring farmer, Jim Ragan, who had plowed the Leopold's garden plot a few days before. At first they dismissed the fire as insignificant, but when Aldo started to sense that it was something more serious, he "became excited and rushed into action" (Meine 2010, 518). Aldo, his wife, and his daughter grabbed some equipment that could be useful in a fire and drove to Ragan's house, where a dozen neighbors had already arrived to fight the fire. He instructed his wife to watch for the fire crossing the road and his

---

1. Curt Meine's (2010) biography of Aldo Leopold describes Aldo's last days; in this section I briefly summarize Meine's more detailed account.

daughter to call the local fire department, urging her to name-drop the Leopolds. Eventually, after first saying that it was too far, the fire department sent one fire truck. Meanwhile mother and daughter continued with the firefighting efforts and "were watching for Aldo, but so many people had gathered by then that they could not locate him" (Meine 2010, 520). Unfortunately, they were to learn that Aldo (who had been in poor health) had suffered a heart attack while carrying a water pump to fight the flames. He was found lying on his back with his arms folded across his chest, dead at age 61.

Meine writes,

> As word of Leopold's death fanned out across the continent, those who had worked with him over the years were shocked into the realization that they had known a remarkable man. They knew it while he was alive, of course, but Leopold had exerted his influence in the conservation movement so skillfully that many had taken his guiding presence for granted. Consolatory letters and telegrams poured in to Madison.... Beneath the expressions of grief, there ran an undertone of gratitude for having been fortunate enough to know Leopold, to work, go afield, and share a conversation with him. (Meine 2010, 521–22)

This abbreviated description of Aldo Leopold's last days reveals some of the overlapping human communities that Leopold was a part of, including the conservation communities with whom he worked and corresponded and the farming community in Sauk County. The story of the Sauk County farming community coming together to fight the fire at Jim Ragan's house makes the interdependence of community members clear (see chapter 2 for further discussion of the concept of interdependence used here, a conception that includes *vulnerability*). The fire had threatened nearby farmhouses and dwellings. This is one type of interdependence, where the physical proximity of the buildings (and the local plants and moisture conditions) gave rise to vulnerabilities. An action that Jim Ragan had taken—starting a trash fire—inadvertently put other members of the community, their shelters, and their livelihoods at risk. The response of his neighbors, including the Leopolds, illustrates another type of interdependence—the *cooperative* efforts taken, even at a risk to themselves. Were these farms in *competition* with each other? Perhaps so, and if so, that would be yet another illustration of interdependence between members of a community, in the sense that there were causal interactions between the farms that could have affected their fates positively or negatively and again given rise to vulnerabilities.

Leopold thought that interdependence was the basic premise on which ethics in general rested, and he believed that, since interdependence manifests not just in human communities like the farming community in Sauk County but also in land (a.k.a. biotic) communities, our ethics should be extended to include land communities as well. The goal of this chapter is to clarify and elaborate this argument of Leopold's for the land ethic and to show that the premises on which it rests are plausible and defensible.

What is the land ethic? Answering that question is, of course, the project of this entire book. It cannot be easily summarized. Yet philosophers like short, snappy summary statements; thus, many have latched on to Leopold's statement that "a thing is right when it tends to preserve the integrity, stability, and beauty of the biotic community. It is wrong when it tends otherwise" (1949, 224–25). As discussed in chapter 1, these sentences have been misinterpreted, and these misinterpretations have spread, giving rise to widespread fictions that Leopold would have advocated sacrificing individuals for the community (in spite of his explicit statements otherwise), or that by "stability" he meant keeping communities unchanged (rather than the capacity of the land for self-renewal that he emphasized throughout "The Land Ethic" and in other essays). With the caveats that Leopold believed that "the land ethic [is] a product of social evolution" and that "nothing so important as an ethic is ever 'written'" (1949, 225), here is my rough stab at a summary statement for the land ethic:

> In addition to the obligations that we already have toward other human individuals and to our human communities, act so as to protect and promote the capacity of land communities (soils, waters, plants, and animals, understood collectively) for self-renewal, i.e., their health, implying respect for both community members and the community as a whole.[2]

---

2. How to balance these sometimes competing rights and obligations is a topic for the next chapter.

And as the epigraph to this chapter clarifies, "A land ethic of course cannot prevent the alteration, management, and use of these 'resources,' but it does affirm their right to continued existence, and, at least in spots, their continued existence in a natural state" (Leopold 1949, 204).

In what follows, I elaborate on Leopold's main argument for the land ethic as well as some other arguments he gave in support of it. I then identify various explicit and implicit premises underlying the land ethic and show that each is at least plausible and reasonable. Note that this chapter, unlike previous chapters, draws mostly on the essay "The Land Ethic" in ASCA, although I draw on other works where relevant. It will also of necessity draw heavily on other chapters of this book, to which the reader is referred for additional discussion. Despite its influence in environmental ethics, the essay "The Land Ethic" and ASCA as a whole was written for a general audience, not a philosophical audience, and Leopold was trained as a scientist (initially, as a forester), not as a philosopher. It is important to read him in that light. Along the same lines, bear in mind that some aspects of his views may simply not be fully or explicitly spelled out, such as his general views on ethics.[3]

LEOPOLD'S MAIN ARGUMENT FOR THE LAND ETHIC

Leopold's main argument for the land ethic rests upon seeing that most people already accept the basic principles on which a land ethic can be justified. That is, once we examine what justifies and grounds our human ethics, and once we understand land communities properly, we should see that the same principles justify and ground extending obligations to the land. The discussion in this section further elaborates and defends this argument, which I think can be understood as an appeal to consistency in our ethical think-

---

3. This is perhaps one of the reasons for varying interpretations: Callicott (1987) reads Leopold as a Humean-type ethicist and Norton (1988, 2005) reads him as an American Pragmatist. See chapter 1 for further discussion. My goal in this chapter is to stick as closely as possible to Leopold's text so as to limit the potential for making inferences that go beyond what he intended.

ing, although Leopold did not phrase it in those terms or make the argument as explicitly as I will make it.

Central to Leopold's main argument for the land ethic is his claim that "all ethics so far evolved rest upon a single premise: that the individual is a member of a community of interdependent parts" (1949, 203). This informs how Leopold thinks of relations and obligations in *human* communities. He tells us that an "ethic, ecologically, is a limitation on freedom of action in the struggle for existence" whereas an "ethic, philosophically, is a differentiation of social from antisocial conduct," but that, really, these "are two definitions of one thing" (1949, 202). And, he said, "the thing has its origin in the tendency of interdependent individuals or groups to evolve modes of co-operation," modes which ecologists call "symbioses" (1949, 202).

But Leopold acknowledged that in human communities, cooperation has not entirely replaced competition; both types of behaviors persist. (As I suggested above, Leopold's own Sauk County community is one such example). And along with those behaviors, we accept limitations on our freedom of action—obligations and rules of social conduct—because we are parts of human communities made up of interdependent and vulnerable members. For example, Leopold noted that "the existence of obligations over and above self-interest is taken for granted in such rural community enterprises as the betterment of roads, schools, churches, and baseball teams" (Leopold 1949, 209). These obligations may in some cases serve our self-interest, but are generally accepted even when they do not: for example, by people without children who recognize that it is in our collective interest to have an educated populace.

However, Leopold argued, our ethical obligations are not limited to humans. Both *history* and *ecology* teach us that it isn't just humans we are interdependent with. We are also interdependent with "soils, waters, plants, and animals, or collectively: the land" (1949, 204).

With respect to history, Leopold argued that "[m]any historical events, hitherto explained solely in terms of human enterprise, were actually biotic interactions between people and land" with the characteristics of the land determining "the facts quite as potently as the characteristics of the men who lived on it" (1949, 205). What if, Leopold asked, when Kentucky was subject to colonization by white settlers, "plow, fire, and ax" had yielded not bluegrass but "some worthless sedge, shrub, or weed" (205)? The outcome, Leopold suggested, might have been very different: "We are commonly told what the human actors in this drama tried to do, but we are seldom told that their success, or the lack of it, hung in large degree on the reaction of particular soils to the impact of the particular forces exerted by their occupancy" (206). Other areas, such as the southwestern United States, were less lucky. These examples illustrate the interdependencies between humans and the rest of the land community, interdependencies that are often overlooked but frequently play an essential role in human flourishing (or lack thereof, again highlighting our vulnerability).

With respect to ecology, an examination of Leopold's conception of food chains likewise reveals interdependencies between all the parts of the land community, including humans. As discussed in chapter 2, Leopold characterized food chains in a broad sense to include not just the dependencies of organisms that feed on each other, but also other dependencies in the form of "services" that organisms furnish for each other, such as shade or shelter.[4] Interactions underlying such interdependencies include both positive and negative ecological interactions, such as predator-prey, parasite-host, competition, and mutualism. Leopold explicitly included humans within these food chains as well as abiotic components, such as soil and water. All parts of the land community are interdependent; some are directly connected via the aforementioned ecological interactions, whereas others are only indirectly connected through the interactions of other parts. An example discussed in chapter 2 illustrates this point, highlighting a chain of interdependencies between farmers, other humans, cows, watercourses, soil, trout, wildflowers, partridges, woodcocks, and more—a case where the seemingly simple act of a farmer clearing a slope ends up affecting all, showing the vulnerability of all. Moreover, within a land community, each part is a link in many such chains, forming a "tangle" of food chains and a web of interdependencies. As dis-

---

4. See chapter 2 for further discussion of the concept of interdependence and defense of the ideas summarized here.

cussed in chapter 3, these webs of interdependencies are the basis of land communities.[5]

With the interdependencies within land communities established, the basic structure of Leopold's argument can be clarified as follows: Given interdependence between humans, we accept limitations on our actions (rules of conduct) to benefit and protect individual humans and human communities; our ethical theories capture these rules of conduct. However, history and ecology show us that we are interdependent with more than other humans; we are also interdependent with other species and with abiotic components such as soils and waters, via both cooperative and competitive interactions. We form land communities together with all of these entities. Thus, consistency demands that we need to extend (expand) our ethics to include the land; by the same logic, we need to accept limitations on our actions (rules of conduct) to benefit specific parts of the land community as well as the community as a whole. The land ethic, Leopold states, "implies respect for [humans'] fellow-members, and also respect for the community as such" (1949, 204). As noted above, Leopold saw our coming to accept the land ethic as a part of an ongoing process of social (i.e., cultural) evolution. Consistent reasoning about our obligations toward the land can be seen as a step in that process.

What does it mean to respect the land community as a whole? Leopold clarified that the land ethic "reflects a conviction of individual responsibility for the health of the land"—understood as "the capacity of the land for self-renewal"—with conservation as "our effort to understand and preserve this capacity" (1949, 221).[6] As discussed in chapter 4, this implies maintaining and promoting species biodiversity by preserving interactions between species and retaining long food chains, as well as maintaining and promoting

---

5. See chapter 3 for further discussion of the concept of land communities and an explanation of how they can be understood as individual biological wholes with boundaries (albeit fuzzy ones).

6. Although Leopold was not solely focused on individual responsibility for the land (vs. societal or governmental responsibility for it), in "The Land Ethic" he emphasized that the actions of individuals with respect to their private property (itself a governmental policy) were an overlooked and essential part of conservation.

soil health. It includes obligations such as "bettering the behavior of the water that falls on the land," or in the preserving of "the beauty or diversity of the farm landscape" (Leopold 1949, 209).

If this interpretation is correct, it puts Leopold's argument in a category with similar arguments from consistency in environmental ethics, such as those of Singer (1979) and Regan (1983), both of whom argue for the similar capacities of human and nonhuman animals as a reason for extending our ethical theories (utilitarian and Kantian, respectively) to nonhuman animals, or else risk an inconsistent or arbitrary ethical theory.[7] The ethical upshots for Singer, Regan, and Leopold are, of course, different; my point here is that the reasoning is the same for each.[8] Although this form of reasoning has its limitations, it is widely accepted among both philosophers and others (Newman et al. 2017; Varner 2020); it may in fact be the most widely accepted approach in ethics, even though it ultimately must rest on an initial premise that is not itself argued for.[9]

OTHER ARGUMENTS FROM LEOPOLD FOR THE LAND ETHIC

Leopold gave other arguments for the land ethic besides the argument for consistency. First and foremost, I think it's fair to say that the entirety of ASCA is an argument (albeit not a deductive one) for the land ethic. Leopold stated that, "It is inconceivable to me that an ethical relation to land can exist without love, respect, and admiration for land, and a high regard for its value" (1949, 223). Surely the earlier chapters in the book, giving us a peek into the lives of mice, skunks, oaks, geese, various flowers, prairie, woods, streams, and much more, are meant to show us why we should love, respect,

---

7. This makes the argument for the land ethic a type of ethical extensionism, where it is argued that an existing ethical theory ought to be extended to include a larger sphere than it traditionally covered.

8. See also my discussion of whether Leopold's argument can be characterized as using the method of reflective equilibrium, which further elaborates the ideas here (Millstein 2020).

9. Nolt (2006) finds the form of the argument valid but thinks that it is only sound when *sentience* is the capacity in question. However, since Nolt provides no reason for his claim about soundness other than his own intuition, we need not consider it further.

and admire the land and have a high regard for its value. Indeed, in the foreword, Leopold stated that he is hoping to effect a "shift of values" (1949, viii). These much beloved pages of ASCA are intended to inspire appreciation for the beauty of the land and the need to conserve it.

Another argument is only hinted at in the essay "The Land Ethic," but its roots can be seen in earlier essays. Not only did Leopold repeatedly refer to plants, animals, soil, and water as "parts" (or "components"), suggesting a whole; he pointed out that even parts that lack economic value may be essential to the healthy functioning of the land community (1949, 214). In 1935, he wrote, "Philosophers have long since claimed that society is an organism, but with few exceptions they have failed to understand that the organism includes the land which is its medium" (Leopold 1935b, 212). In 1938, he further elaborated this idea: "Harmony with land is like harmony with a friend; you cannot cherish his right hand and chop off his left. That is to say, you cannot love game and hate predators; you cannot conserve the waters and waste the ranges; you cannot build the forest and mine the farm. The land is one organism" (Leopold 1938b, 416).

However, these holistic expressions do not mean that Leopold thought we should *only* consider the whole, as some philosophers have contended concerning Leopold (e.g., Regan 1983). On this point, Leopold wrote, "If the components of land have a collective as well as a separate welfare, then conservation must deal with them *collectively as well as separately*. Land use cannot be good if it conserves one component and injures another. Thus a farmer who conserves his soil but drains his marsh, grazes his woodlot, and extinguishes the native fauna and flora is not practicing conservation in the ecological sense. He is merely conserving one component of land at the expense of another" (Leopold 1944a, 316; emphasis added).

From these passages I glean the following (implicit) holistic argument: Humans have obligations to soils, waters, plants, and animals, but it would be a mistake to recognize *only* these obligations. These are all parts of a larger whole, and the welfare of the parts cannot be separated from the welfare of the whole. I understand this to be a metaphysical claim, not a pragmatic one; he's not saying that it is in humanity's self-interest to protect the land community (although that may also be true), but that favoring some of the parts over others (whether those parts are humans or deer or soil or any other part) is as nonsensical as cherishing a right hand and chopping off the left.[10] Thus, if you accept obligations to any soils, waters, plants, or animals (including humans) already, you are implicitly committed to the whole as well. The parts cannot be separated from the whole, and because of interdependencies, attempts to favor any particular part over others will eventually backfire.

Finally, Leopold also suggested that humans simply aren't smart enough to pick and choose between parts of a community or to play the conqueror—that to do so is eventually "self-defeating." He states that "it is implicit in such a role that the conqueror knows, *ex cathedra*, just what makes the community clock tick, and just what and who is valuable, and what and who is worthless, in community life. It always turns out that he knows neither, and this is why his conquests eventually defeat themselves" (1949, 204).

According to Leopold, this ignorance extends both to human communities and the larger communities that contain them because, in the case of land communities, the "biotic mechanism is so complex that its workings may never be fully understood" (1949, 205).[11] I suspect that here he had in mind situations such as his earlier embrace of predator eradication in an attempt to increase the numbers of deer that could be hunted—an attempt that did indeed turn out to be self-defeating: killing all the predators caused the deer population numbers to skyrocket, eventually leading to the destruction of local plant life and the starvation of deer (see chapter 2 for further discussion). So, according to Leopold, we can't be conquerors, we can't choose which species to favor over other species, because we don't understand the complexity of interdepen-

---

10. Of course, there are cases when we might sacrifice a hand, but these are extreme cases (e.g., when gangrene threatens the entire body). In such extreme cases, where there is no other alternative, Leopold might likewise have endorsed the removal of a member of a land community.

11. Today we still struggle with our "picking and choosing," sometimes finding that attempts to eradicate invasive species have a rebound effect; see, for example, Zhao et al. 2020 and Grosholz et al. 2021. That's not to say that Leopold would be categorically against attempts to eradicate invasive species, just that we need to proceed cautiously and recognize that we are likely to make mistakes, so (as suggested in the previous footnote) we might attempt this only in the most extreme of circumstances.

dencies in land communities. Leopold asks, "Who but a fool would discard seemingly useless parts?" since "[keeping] every cog and wheel is the first precaution of intelligent tinkering" (1938b, 417). Since we can't pick and choose, it follows that we have to recognize our obligations to all species and to the land community as a whole.

Although I do not directly discuss these three arguments further, I do think they play an important role in Leopold's thinking, and elements of them (particularly the argument from ignorance and the holism argument) can help to justify his main argument, as I show below.

DEFENDING LEOPOLD'S PREMISES

Having acknowledged these other arguments for the land ethic, in this section I turn back to what I have suggested is Leopold's main argument, the argument from consistency. Here I defend that argument by trying to identify what I take to be premises of Leopold's argument that might be in need of further defense and by providing at least the beginnings of those defenses. In what follows, I use the term *premises* loosely; I don't mean to imply I have provided a deductive reconstruction of Leopold's argument, a strategy that I think would be unhelpful here (again recalling that Leopold was not a philosopher and that ASCA was written for a general audience). Instead, what I am calling premises might just as well be called implicit (or sometimes more or less explicit) assumptions, necessary conditions, or points likely to be subject to counter-arguments. I have attempted to provide a logical flow through these premises; the reader should not take these to be ordered premises within a deductive argument.

Extent of Interdependence

In order for Leopold's argument from consistency to succeed, interdependence would need to be *pervasive* in the land community, not just something that obtains between a few species. That is, we need reason to accept the truth of the implicit premise that all parts of a land community—human, nonhuman, soil, water—are interdependent. This premise was discussed in detail in chapter 2, but I summarize some of the relevant points here.

Some parts of the land community are directly connected via ecological interactions, whereas others are only indirectly connected through the interactions of other parts. Indirect connection via a chain of direct connections is sufficient for interdependence because actions that affect entities at one end of the chain will affect entities at the other end. Thus, interdependence does not require direct connections between all of the parts. Also, each part is a link in many chains, forming a "tangle" of chains.

Consider, for example, the bee "link": A huge percentage of crops and native flowers depend on various bee species for pollination—in turn, human and nonhuman animals depend on those for food. Yet, many bee species are threatened by human-caused global climate changes (Le Conte and Navajas 2008; Soroye et al. 2020) and pesticide use (Holder et al. 2018; Raine 2018). Thus, in this chain, humans and other animals are interdependent with bees, various plant species, and soil and water; all of these parts are interdependent. In particular, note that bees can be harmed or benefited by human actions; they (and we) are *vulnerable*, one aspect of the sense of interdependence that Leopold seems to be deploying (again, see chapter 2 for further discussion). This being just one of many such (interconnected) examples, interdependence is indeed pervasive within land communities.

Moral Considerability of Land Communities

It also seems that Leopold is assuming that land communities are deserving of *direct* moral consideration in moral decision-making. Direct moral consideration means that an entity has direct moral standing because of the sort of entity it is and not simply because of how it might benefit other entities of value, which would be *indirect* moral standing. For example, let's suppose (as most would agree, I think) that horses are deserving of moral consideration. Are they deserving of moral consideration because they are sentient, thinking beings? Or is it only because they are a benefit to their owners, who would be harmed if their horses were harmed? If the answer to the first question is yes, then horses have direct moral standing, based on capacities that *they* (the horses) have; if the answer to the second question is yes, then horses have indirect moral standing, based on the connection between human owners and their horses.

Thus it is possible for an entity to deserve both direct and indirect moral consideration; it is also possible that they deserve only indirect moral consideration. My suggestion in this section is that Leopold saw land as deserving of both direct and indirect moral consideration.

A healthy land community is often a benefit to its members—making it worthy of indirect moral consideration—yet Leopold seems to be identifying the good of a land community as something in addition to those benefits. But *are* land communities directly morally considerable? (In what follows, I just say "morally considerable" and drop the word "direct" for ease of exposition). In an influential paper, Kenneth Goodpaster argues that an entity can only be said to be morally considerable—that is, deserving of direct moral respect, part of the moral sphere—if it has interests (Goodpaster 1978).

Having interests, and moral considerability more generally, might seem to require that land communities not be arbitrarily bounded, so that there is an identifiable, non-stipulative, not-by-convention entity to which our obligations attach. For example, using Yellowstone National Park as a proxy for land community boundaries would be to use a purely stipulative, ecologically arbitrary boundary; treating it as a genuine land community has in practice meant that wolves who wander out of the park lack protections and have been shot by hunters. I take no stand on whether having boundaries is required for moral considerability, but others have (e.g., Newman et al. 2017). This point is discussed in detail in chapter 3, where I argue that Leopold's land community concept (a blend of ecosystem and ecological community concepts) posits an entity that is bounded by large differentials in the flow of material and energy and/or by large differentials in interactions between populations of different species. Lake communities are a clear-cut example where boundaries can be identified, with the insects, fish, water, microorganisms, and so forth, interacting at a higher rate and with a greater matter/energy flow than with entities outside the geographic boundaries of the lake. (See chapter 3 for discussion of less clear-cut cases, including how migratory animals such as birds can be taken into account).

Moral considerability might also seem to require that land communities be entities that can be benefited or harmed or that they have a good or bad of their own—that they have interests. In chapter 4, I argue that *land health*—the capacity of the land for self-renewal, allowing it to persist over time—is a property of land communities that gives them interests (see chapter 4 for fuller discussion). Leopold argued that land health is affected by practices like killing off predators (which has downstream effects on the numbers of prey and everything they interact with) or overgrazing (which can lead to soil erosion and loss of soil nutrients that support life). This is a property of the land community that is somewhat independent of the health of the component parts, since it is a property of the interactions and interdependencies between the members of the land community, and since parts can be unhealthy while the whole is healthy (and vice versa).

With a strong case for the coherence of land community boundaries and a strong case that land communities can be benefited or harmed (by improving or impairing their land health), a strong case can be made that land communities are morally considerable.

Ethical Basis of the Land Ethic

Another premise that is in need of defense in Leopold's main argument is his explicit claim that "All ethics so far evolved rest upon a single premise: that the individual is a member of a community of interdependent parts" (1949, 203). Can interdependence serve as the basis of an ethic? It is beyond the scope of this book to answer this question, so I just note that some precedents might be found, for example in some communitarian ideas (see Bell 2020 for discussion of communitarianism).[12] For example, Charles Taylor, citing Aristotle's *Politics*, writes, "Man is a social animal, indeed a political animal, because he is not self-sufficient alone, and in an important sense is not self-sufficient outside a polis" (1985, 189). Indeed, Leopold's account of the historical evolution of ethics echoes Aristotle's.[13]

---

12. I am using the term *communitarian* as an imperfect ballpark label for views that emphasize or prioritize the community rather than, or in addition to, the members of the community.

13. To be clear, I am not claiming that Leopold was directly or indirectly influenced by Aristotle, since I don't have the evidence to make that claim; I am just noting the striking parallel.

Aristotle's "quasi-historical" account had individuals combining into families and groups and then these groups combining to form a "complete community," the self-sufficient city-state (Miller 2017). Furthermore, Aristotle believed that "the city-state is naturally prior to the individuals, because individuals cannot perform their natural functions apart from the city-state, since they are not self-sufficient" (Miller 2017). The community in the form of a city-state is self-sufficient, on Aristotle's view, but individuals are not. This Aristotelian account seems akin to Leopold's view that, because of interdependence between humans, the first ethics dealt with the relation between individuals, with "later accretions" dealing with the relation between the individual and society (Leopold 1949, 202–3). Of course, that philosophers such as Aristotle and Taylor articulated similar ideas does not make them correct, but it at least serves to show that they have a certain plausibility.

It seems to me that similar views may also be found in the relational ethics of some feminist philosophers, the African idea of ubuntu, and some Indigenous views; see Meynell and Paron (2021) for discussion of these traditions.[14] Or as Martin Luther King Jr. wrote, "Injustice anywhere is a threat to justice everywhere. We are caught in an inescapable network of mutuality, tied in a single garment of destiny. Whatever affects one directly affects all indirectly" (King 1963). Perhaps (although I speculate here) King's sentiment is at the basis of all such views: that, as members of a community, our fates are intertwined, so that interdependence and vulnerability are fundamental aspects of our existence and are thus the origin of our obligations to one another, an aspect that must not be overlooked.

Kinds of Interdependence

Another Leopoldian premise that is in need of defense is the seemingly implied claim that the interdependencies between soil, water, plants, and animals are instances of the same kind as interdepen-

---

14. Robin Kimmerer's (2013) discussion of gratitude and reciprocity toward our nonhuman relatives might be particularly relevant, given her ecological approach—but see Whyte (2015), who cautions against purely abstract comparisons like this that overlook considerable differences between colonial settler approaches and Indigenous ones.

dencies between humans (but perhaps this is not implied—see next subsection), and so ground our obligations to the land, just as our interdependencies with other humans ground our obligations to them. As a starting point toward addressing that issue, consider the story that began this chapter. Perhaps it is the interdependencies within the community of farmers that draw our attention first: the neighbors, their farms, their willingness to help Jim Ragan, the fire truck that is intended to serve the community. As noted above, this story illustrates both how cooperative and competitive interactions gave rise to vulnerabilities, and thus interdependencies, between the human members of the community.[15] As it is a story of livelihoods, survival, and risk, it seems reasonable to assert that the cooperative and competitive interactions in this story are just different manifestations of competitive and cooperative interactions that occur in a land community. That is, they are all ecological interactions, differing in form to be sure, but not in kind; every organism has a story about its livelihood, its survival, its risks. So considering this example, the claim that human interdependencies and interdependencies within a land community are instances of (or different manifestations of) the same kind has some initial plausibility.[16]

More generally, many combined human and nonhuman ecological interactions have to do with survival: competitive interactions, predator-prey interactions, parasite-host interactions, mutualistic interactions. But, at the risk of being reductionistic, many purely human interactions do, too. The most obvious example is paren-

---

15. None of this is to deny that members of the community were surely motivated to help Ragan himself (i.e., the individual). In this story, at least, that motivation is fully compatible with interdependent actions that also preserve the community. In other situations, what is best for the individual might come into conflict with what is best for the community. Such issues are discussed further in the next chapter.

16. I was originally understanding this implicit premise as Leopold making an *analogy* between human-only interdependencies and the interdependencies typically identified by ecologists. But at the suggestion of several helpful commenters, I realized that understanding Leopold's argument to be about recognizing land community interdependencies as instances of the same kind is both more defensible (analogies always being fraught with determining relevant similarities and significant differences), and, I think, closer to the text. On the latter point, for example, Leopold said that "[p]olitics and economics are advanced symbioses" (Leopold 1949, 202).

tal care for an offspring (or vice versa as one gets older), but the truth is that we humans all depend on each other in a variety of ways, whether for providing food, shelter, or transportation—all of which are ecological interactions. In chapter 2, I argued that even interactions that are negative (i.e., in some sense harmful to some entity involved) give rise to interdependencies, and these are likewise present in human societies. To give one example, we have an economic system that is based on competition (which may yield winners and losers), but other areas are arguably based on competition as well, such as education, government, and sports. And, like combined human and nonhuman ecological interactions, although our negative human interactions like competition can have harmful effects in some contexts or on some individuals, they can also have positive effects in other contexts or on other individuals, with interdependencies best understood in the context of a web rather than solely in terms of pair-wise interactions.

Of course, humans are interdependent in psychological ways, too, and these may not occur in *some* species—but they *do* occur in others (e.g., many mammals and birds). In fact, interdependencies are quite varied among species, so that even though we can capture many of the basic types of interactions that tend to give rise to them (again, predator-prey, parasite-host, etc.), they will manifest differently, sometimes very differently, between different species. So the fact that humans might have some unique interactions does not mean that those interactions do not give rise to ecological interdependencies that are of the same general kind as the more traditional sort of ecological interdependencies identified by ecologists.

However, some might wish to press the point that interdependencies that apply only to humans differ importantly from those that apply to both humans and nonhumans, as perhaps the above discussion of Leopold's intellectual communities suggests (although even there, the focus on conservation is arguably also related to survival). As Charles Taylor notes, it is "not just that [humans] cannot physically survive alone, but much more that they only develop their characteristically human capacities in society" (Taylor 1985, 190–91). So, perhaps it can be argued that the social relations between humans, and the benefits of a community qua city-state, are the primary sorts of interdependencies that matter for ethics. But what determines what is primary? Aristotle's own argument in *Politics* maintains that "the city-state is naturally prior to the individuals, because individuals cannot perform their natural functions apart from the city-state, since they are not self-sufficient" (1253a18–29; quoted in Miller 2017). By that reasoning, humans cannot perform their natural functions apart from their land communities, regardless of the "health" of their city-state, which would make a person's land community prior to both the human community as well as the individual. Such a response would seem to embrace holism, echoing the holism discussed above, emphasizing Aristotle's point that the whole is prior to the part, so that if (for example) a whole body were to be destroyed, there would be no foot or hand.

But perhaps such a response goes too far, threatening to sacrifice the individual to the whole, something that Leopold did not endorse. Perhaps it is simply unclear whether the individual or the whole is "prior," and that, although the interdependencies between humans and other humans and between members of a land community might have important differences, they share much in common, often relating to survival (understood broadly to include reproduction) in some way. That is, we can acknowledge a claim like Gary Varner's, which holds that "the value attributed to human communities in modern, pluralist democracies is better understood as based on the way political systems can sustain (or fail to sustain) the rationality and autonomy of humans who understand each other as mutually respecting citizens and holders of legal rights, rather than just sustaining them biologically" (Varner 2020, 15). But this might just be to say that human communities have a social/political value over and above that which is imparted by their biological/survival value. And here it is worth noting that many nonhuman species have social interdependencies as well, even if they turn out not have the same complexity or nuance that human social interdependencies do. One reasonable way of handling this would be simply to say that survival-based interdependencies give rise to a different set of obligations than the more socially oriented ones do (to the extent that the distinction can be drawn—I think there will be overlap). Then the question would become one of how to handle competing obligations, a topic that is discussed in the next chapter. This would permit us to recognize the legitimacy of seeing all interdependencies, whether of humans only or of both humans and nonhumans, as instances of the same kind, while not dismissing any important differences.

Interdependence Is Needed for Completeness

But there is another way of thinking about the Sauk County story. A closer look reveals that it is not just a story about humans; humans are just one part of the story, perhaps not even the most important part. The Leopolds were there to plant trees and to enjoy the local wildlife, and they are surrounded by farmers whose livelihoods depend on growing crops, activities that all depend on the soil and water in turn. Moreover, the entire story of putting out the fire— how quickly it spread and how quickly it was extinguished—was dependent on the surrounding brush, crops, and marshland; their levels of moisture; and the availability of water to douse the fire. The interdependencies between the members of the human community were only some of the relevant interdependencies within the larger land community in Sauk County in 1948 that the Leopolds were part of. The interdependencies between humans, other animals, plants, soil, and water are the interdependencies that form the complete web. So perhaps this is not really a question of understanding different interdependencies as instances of the same kind, as proposed in the previous subsection; perhaps it is a question of recognizing all of the actors in a drama, as Leopold himself suggests in "The Land Ethic."

This way of understanding and defending Leopold's consistency argument has a bit of a holistic cast to it, but does not suggest that the land community is an indivisible whole, akin to an organism. Rather, it simply points out that human interdependencies alone never tell the full story. Thus, a more accurate and more complete accounting of the relevant interdependencies must always include nonhuman organisms, soils, and waters, from which it follows—if ethics is grounded in interdependencies—that we ought to extend our ethics from the human realm to include the rest of the land community. Indeed, Leopold explicitly argued for the study of the nonhuman and the human together: "Land ecology is putting the sciences and arts together for the purposes of understanding our environment.... *Land ecology discards at the outset the fallacious notion that the wild community is one thing, the human community another.*

What are the sciences? Only categories for thinking. Sciences can be taught separately, but they can't be used separately, either for seeing land or for doing anything with it" (Leopold 1942b, 302–3; emphasis added).[17]

In the same essay, Leopold deploys a drawing (see fig. 2.1) that depicts the lines of dependency in a community. Rock, soil, alfalfa, cow, farmer, grocer, lawyer, and student all appear on the same "food chain" (in that order), along with the overlapping areas of study needed to examine them (geology, botany, agronomy, animal husbandry, sociology, and economics). The relevant point here would be that, just as the sciences should jointly study the land community and *all* of its interdependencies, so should ethics take into account the most accurate and complete picture of the morally considerable entity. Any ethics that failed to do that would likewise be incomplete. Thus, even if the argument concerning interdependencies as a general kind fails—even if it turned out that human and nonhuman interdependencies were very different—there is still a case for extending our moral obligations to the entire land community. The consistent position for someone who already accepted our obligations to humans and human communities would be to accept obligations to our land communities as well.

It may also be that Leopold is making *both* arguments: that interdependence in a land community is the same kind of thing as human interdependence *and* that interdependence within the land community is required for a complete account of our moral obligations. After all, the two arguments are fully compatible. In any case, I hope to have provided some reason to accept both of them.

Intrinsic Value

Finally, Leopold seems to imply a premise that land communities are of intrinsic value; he famously stated in "The Land Ethic" that the value of the land is "something far broader than mere economic value; I mean value in the philosophical sense" (Leopold 1949, 223). Along these lines, the limitations of an economic approach are a

---

17. See Meine 2020 and Van Auken 2020 for further discussion of Leopold's desire to integrate social and ecological analyses.

strong theme in the essay, noting, for example that a "basic weakness in a conservation system based wholly on economic motives is that most members of the land community have no economic value" (210) and that a system of conservation based solely on economic self-interest "assumes, falsely, I think, that the economic parts of the biotic clock will function without the uneconomic parts" (214). Again, interdependence is key here, with economic parts of the land community dependent on uneconomic parts (and vice versa). So, if one thinks that the land has *some* sort of value (which seems like a minimal assumption), and then learns that a purely economic approach is too limited and bound to fail (i.e., it is self-defeating), it's reasonable to think that the land has "value in the philosophical sense" (i.e., intrinsic value).[18] Another way to put the point is to say that economic value is not a viable, workable guide to action, given that it does not allow one to preserve the very things that one values (because it is self-defeating).

The logic of the rejection of economic value alone as a viable understanding of the value of a land community can, I think, reasonably be deployed to reject instrumental value as a sufficient understanding of the value of a land community as well. Just as it might seem as though some members of the land community have no economic value, it might similarly seem as though some members of the land community have no instrumental value and can thus be disregarded, and yet, given our ignorance about what makes the land community tick (as discussed above), those choices would eventually be self-defeating. The "weak anthropocentrist" (Norton 1984) might argue that we need only act as if our land communities have value, while still acting from the motive of human obligations alone, but the jury is still out as to whether we have the ability and the knowledge to make the choices that such mental gymnastics would require (Westra 1997).

To be clear, Leopold was neither denying that the land and its components have economic and instrumental value, nor was he denying that we should act on the basis of those values. He stated this explicitly: "It of course goes without saying that economic fea-

---

18. Callicott (1987) also sees Leopold as maintaining that the biotic community has intrinsic value; on this point, we agree. See Norton (2005, 2011) for an alternate view.

sibility limits the tether of what can or cannot be done for land. It always has and it always will. The fallacy the economic determinists have tied around our collective neck, and which we now need to cast off, is the belief that economics determines *all* land-use" (Leopold 1949, 225; emphasis in original). His point is that economic and instrumental value *alone* are insufficient and self-defeating. Thus, *in addition* we must recognize that the land has intrinsic value. Things that have value in and of themselves (intrinsic value) can also have economic and instrumental value (value to us as human beings)—one's friends are an obvious example. Leopold is acknowledging that we will use the land to our own benefit, but we must do so in a way that takes into account and recognizes the value of the whole land community.

In a Leopoldian spirit, we might add an additional consistency argument to make the case for the intrinsic value of land communities. Leopold said, "There are two things that interest me: the relation of people to each other, and the relation of people to land" (quoted in Meine 2010, 51). Thus, one might argue that if human communities have intrinsic value in virtue of the sorts of entities that they are (webs of interdependent humans—perhaps parasitic on human value or perhaps as the most reasonable whole), then there is reason to think that land communities also have intrinsic value in virtue of the sorts of entities that they are (webs of interdependent organisms along with soil, water, etc.). But more would need to be done to spell that out thoroughly—why interdependence gives rise to human and land communities with intrinsic value.[19]

As for what sense of intrinsic value Leopold intended, I think that is simply unclear. Environmental ethicists have distinguished between *subjective intrinsic value*, where "intrinsic value is created by human valuing" and "something has intrinsic value if it is valued

---

19. Here a concern might arise regarding human communities that we would consider ethically problematic, such as a white supremacist community. But it is important to recall that intrinsic value, as we usually understand it, is independent of such judgments. Those who hold that humans have intrinsic value typically believe that *all* humans are intrinsically valuable, even though we might condemn some of them for their unethical behavior. The same would be true for our judgments about human (and land) communities.

for what it is, rather than for what it can bring about," and *objective intrinsic value*, where something "has properties or features in virtue of which it is valuable, independent of anyone's attitudes or judgments" (Sandler 2012). On the latter view, intrinsic value is something that humans discover rather than create. Frankly, I am not sure that the distinction matters for the issues at hand, and I think either notion of value can be attributed to Leopold. The point is simply that land communities are intrinsically valuable because of the sorts of entities that they are—webs of interdependent plants, animals, soils, and waters that, when functioning via matter and energy flow through the interactions of the component parts, can support the capacity of the land to self-renew (i.e., land health).

CONCLUSION

I hope to have shown that Leopold's main argument for the land ethic, which I have called an argument from consistency, is reasonable, plausible, and based on defensible premises. Arguments from consistency are a well accepted and straightforward form of argument that can compel us to accept what we might otherwise resist. By appealing to something we already accept—our obligations to our human communities—we can find clear and compelling reasons to recognize our obligations to the land communities that we are a part of. With the ongoing climate crisis, rapid extinction of species, and loss of habitat, we need more than ever to understand that we cannot just focus on ourselves without recognizing all the biotic and abiotic entities with which we are interdependent.

But how can we put Leopold's land ethic into practice? How can it form the basis for policy, especially in situations where what is best for individual members of the land community conflicts with what is best for the land community itself? These are topics for the next chapter.

CHAPTER SIX

Policy Implications

There are two ways to apply conservation to land. One is to superimpose some particular practice upon the pre-existing system of land-use, without regard to how it fits or what it does to or for other interests involved.

The other is to reorganize and gear up the farming, forestry, game cropping, erosion control, scenery, or whatever values may be involved so that they collectively comprise a harmonious balanced system of land-use.

ALDO LEOPOLD, "Coon Valley: An Adventure in Cooperative Conservation"

INTRODUCTION: COON VALLEY EROSION PROJECT

In 1933, Leopold was appointed as chair of game management at the University of Wisconsin, a position initially established within the Department of Agricultural Economics (Meine 1987). The position entailed that he serve as a wildlife extension specialist, and it was in this capacity that he became an adviser to the Coon Valley Erosion Project (Meine 1987). Coon Valley is located in the Coon Creek watershed in southwestern Wisconsin; a new federal agency, the Soil Erosion Service (later to become the Soil Conservation Service), had "selected Coon Creek as the first watershed in which to demonstrate the values of soil conservation measures" (Helms 1992, 51). The goal was to show "how farmers could plan farming operations to include soil conservation for long-term productivity" (51). At the outset of the project, Leopold wryly noted that Coon Valley was "one of the thousand farm communities which, through the abuse of its originally rich soil, has not only filled the national dinner pail, but has created the Mississippi flood problem, the navigation problem, the overproduction problem, and the problem of its own future continuity" (1935a, 49). As Helms described it, "Most of the area was beset by erosion problems. Gullies hindered farming. Coon Creek was subject to frequent, intense floods. Some valuable bottom land had reverted from cropland to pasture due to floods. Trout abandoned the sediment clogged stream" (Helms 1992, 51).

What could have happened, Leopold suggested, was that "some one group would prescribe its particular control technique as the panacea for all the ills of the soil" (1935a, 48–49). The groups that might do so included one "that would save land by building concrete check dams in gullies, another by terracing fields, another by

POLICY IMPLICATIONS

planting alfalfa or clover, another by planting slopes in alternating strips following the contour, another by curbing cows and sheep, another by planting trees" (49). But the federal Soil Erosion Service didn't choose one method over the others; it used them all, "to its lasting credit," Leopold said (49).

From the outset, the Soil Erosion Service also recognized "that sound soil conservation implied not merely erosion control, but also the integration of all land crops. Hence, after selecting certain demonstration areas on which to concentrate its work, it offered to each farmer on each area the cooperation of the government in installing on his farm a reorganized system of land use, in which not only soil conservation and agriculture, but *also forestry, game, fish, fur, flood control, scenery, songbirds, or any other pertinent interest were to be duly integrated*" (Leopold 1935a, 49; emphasis added). Leopold thought that the Coon Valley Erosion Project, by bringing together all of these diverse practices, exemplified what he called "the Principle of Integration of Land Uses" (48). In his time, this sort of integration had been done to some extent with national forests on public land; now, it would be done with agriculture and on private land. In Leopold's opinion, "each of the various public interests in land is better off when all cooperate than when all compete with each other" (48). He hoped that the project would show that "integration is mutually advantageous to both the owner and the public" (48).

As for how this Principle of Integration of Land Uses could be put into practice, Leopold seemed to find the nightly "bull sessions" (his term) to be particularly valuable: "One may hear a forester expounding to an engineer the basic theory of how organic matter in the soil decreases the per cent of run-off; an economist holds forth on tax rebates as a means to get farmers to install their own erosion control. Underneath the facetious conversation one detects a vein of thought—an attitude toward the common enterprise—which is strangely reminiscent of the early days of the Forest Service" (1935a, 54).

This illustrates how Leopold believed that many different interests, values, and perspectives could come to the table and yet share an attitude toward the common enterprise.[1] This shared attitude

---

1. As the epigraph illustrates, Leopold spoke of multiple "interests" as well as multiple "values." In what follows, I do the same, not because I think interests and values are the same, but because I think there are generally values implicit in any

toward the common enterprise, a common purpose supporting the goal of a harmonious balanced system of land use, is identified in the epigraph to this chapter, which is quoted from Leopold's discussion of the Coon Valley Project. Note that Leopold recognized that we "shall never achieve harmony with land, any more than we shall achieve justice or liberty for people" and that "the important thing is not to achieve, but to strive" (1938b, 423). Thus, according to Leopold, seeking harmony in land use is an ideal that we should strive for.[2]

In the short term, Leopold saw successes, such as when "the population of quail in 1934–35 was double that of 1933–34, and the pheasant population was quadrupled," as well as "disappointments and mistakes," such as when a "December blizzard flattened out most of the food-patches and forced recourse to hopper feeders" or when "willow cuttings planted on stream banks proved to be the wrong species and refused to grow" (Leopold 1935a, 53). In the longer term, the project came to be seen as successful (see, e.g., Meine and Nabhan 2014; Meine 2017). Helms writes,

> Since Coon Valley is one of the nation's most studied watersheds, we know the effects of the conservation practices on erosion and sedimentation of streams. In a 1982 study, Stanley W. Trimble, geographer at the University of California at Los Angeles, and Steven W. Lund, U. S. Army Corps of Engineers, used earlier sedimentation studies by Vincent McKelvey and Stanford Happ in assessing the current situation. They calculated that erosion has been reduced at least 75 percent since 1934. Sediment reduction came without converting much cropland to other uses. There has

---

"interest" that is expressed. For example, a forester might value trees, the human uses to which trees are put, and/or the forests of which trees are a part. As discussed in this chapter, Leopold seemed to want to be quite inclusive in the values that would enter into any decision that would affect the environment.

2. According to Meine, "[t]he aim was not to reestablish a vanquished ecological community and its attendant species composition"; instead, "[p]riority was given to immediate problem-solving, enhancing economic and ecological resilience, and coordinating conservation aims" (Meine 2017, 222). So this was not a "restoration" in the sense of restoring exactly what had been there before, but in the broader sense of restoring healthy socio-ecological functioning. Perhaps a more appropriate term would be *rehabilitation*.

been a 6 percent reduction in cropland since 1934. With less sediment flowing into Coon Valley, the trout returned as Raymond Davis had hoped and expected. (Helms 1992, 53)

Leopold, however, seemed discouraged with the longer-term success of the project; in "The Land Ethic," he decried farmers who had been offered Civilian Conservation Corps labor, machinery, and materials in 1933 but "continued only those practices that yielded an immediate and visible economic gain for themselves" (Leopold 1949, 208). This suggests that, in Leopold's mind, implementing the Principle of Integration of Land Uses also required a change in values and a recognition of obligations to the land—an obligation to promote land health—over and above self-interest.

In what follows, I infer eight subprinciples of conservation policy based on some of Leopold's policy-related activities and his stated reflections about them in an attempt to spell out the Principle of Integration of Land Uses.[3] I then clarify and elaborate what I take to be the two biggest challenges of implementing these subprinciples: the challenge of balancing values and interests, and the challenge of incorporating land health into policy. I then address some other potential issues of concern.

EIGHT LEOPOLDIAN SUBPRINCIPLES OF CONSERVATION POLICY

From Leopold's reflections on the Coon Valley Erosion Project, we can glean the first five subprinciples of the Principle of Integration of Land Uses—that is, a Leopoldian conservation policy:

1. Include and attempt to integrate all pertinent interests and values.
2. Seek cooperation rather than competition between the different interests and values to try to find a harmonious, balanced system of land use.

---

3. To be clear, although I think these inferences are justified, they are in the end my inferences, not something that Leopold clearly stated as, "This is how we should do conservation policy." Thus, I may have failed to include some subprinciples that he might have included were he to have made them explicit, or I may have inadvertently introduced other misrepresentations.

3. Deploy a variety of techniques.
4. Recognize that there will be failures, some due to ignorance and some due to unforeseen circumstances (vagaries of weather, etc.).
5. Recognize and act on obligations to the land over and above self-interest, in particular, obligations to promote and protect the health of the land.

The first subprinciple should be understood very broadly, as the discussion of the project makes clear—it includes not only interests and values like agriculture and forestry but also aesthetics and "songbirds." As Leopold said, it should be understood to include whatever values may be involved, any pertinent interest. In this sense, Leopold is endorsing a pluralistic approach to conservation policy. On this point, Curt Meine notes, "The arc of Leopold's career clearly shows him moving away from the top-down and expert-driven approach to land and resource management that marked the early Progressive conservation movement, and toward ever more democratic and participatory land conservation processes" (Meine 2022, 175). The second subprinciple, trying to find a harmonious and balanced system of land use, raises some obvious and difficult challenges for implementing that pluralistic approach; this is discussed further below. The third subprinciple, deploying a variety of techniques, is part of incorporating a variety of interests, given that different techniques may come from different interests, but it also serves to recognize that not all approaches may be successful. The fourth subprinciple is one that Leopold was quite personally aware of, given the repeated attempts of his family to restore trees to their property in rural Wisconsin; it implies perseverance. Finally, the fifth subprinciple is the manifestation of the land ethic. When it is conjoined with the first subprinciple, we can see that protecting and promoting land health (see chapter 4 for a discussion of land health) should be seen as a common goal of all interests (see chapter 5 for a discussion of the arguments for the land ethic). That is not to say that protecting and promoting land health necessarily overrides those interests, as I have mentioned several times. This too is discussed further below.

Agriculture was just one of the areas to which Leopold turned his policy efforts. He was also concerned with forestry and game management, among other areas, so here I detour for a moment to draw out some additional policy subprinciples that are not as obvious from the discussion of Coon Valley. His efforts concerning deer herds on the Wisconsin Conservation Commission, a commission that he served on from 1943 until his death in 1948, are particularly instructive. Time and time again, Leopold recommended policies that would have reduced the size of the deer herd by allowing does to be killed,[4] but except for a highly criticized 1943 hunting season, he was repeatedly outvoted by his fellow commission members (Meine 2010). This is in spite of the fact that "Leopold and the other commissioners worked well together on the wide variety of other issues before them, everything from warden pensions and ice-fishing seasons to tractor purchases and state park concessions" (Meine 2010, 488).

As with soil conservation practices, Leopold recommended incorporating perspectives from varied interests: "The Commission needs the combined judgement of the technical deer men, the wardens, the rangers, the foresters, and the sportsmen on these difficult questions of local status" (Leopold 1944c; quoted in Meine 2010, 462). He also urged the commission to take "the long view":

> This Commission was created, and was given regulatory powers, for the express purpose of insulating it, to some degree, from the domination of fluctuating public opinion. It was hoped that such a Commission *might take the long view, rather than the short view, of conservation problems.* I cannot escape the conviction that if we fail to reduce the deer herd now, we are taking the short view.... My plea is that we vote on this issue, not as delegates representing a County, but as *statesmen representing the long view of Wisconsin as a community.* (Leopold 1946a; quoted in Meine 2010, 488; emphasis added)

---

4. Not that Leopold thought that human hunting of deer was the best method of controlling the deer herd: "It is all very well, in theory, to say that guns will regulate the deer, but no state has ever succeeded in regulating its deer herd satisfactorily by guns alone. Open seasons are a crude instrument, and usually kill either too many deer or too few. The wolf is by comparison, a precision instrument; he regulates not only the number, but the distribution, of deer. In thickly settled counties we cannot have wolves, but in parts of the north we can and should" (Leopold 1944c; quoted in Meine 2010, 456).

The short view allowed the hunt to go forward as usual with the fewest complaints from the public. The long view, however, recognized that failure to reduce the deer herd would mean deer starvation when the next hard winter came, a sacrifice of the future deer herd, and damage to the forest; that is, harm to individual deer, the deer herd, and the community more generally. Leopold later referred to this as "mortgaging the future" (Leopold 1947a).

Leopold suggested that it was "perhaps natural that these risks should loom larger to a forester and game ecologist" like himself than they would for others, and he referenced his almost-finished study of "about a hundred deer irruptions in other states" (Leopold 1946a, 3). Moreover, he doubted that "anyone but a forester can fully visualize this process by which excess deer gradually pull down the quality of the forest" (Leopold 1946a, 3). Thus, in coming to his policy recommendations, Leopold drew on both his former career as a forester as well as his current scientific findings. But Leopold recognized that fear of public criticism was the main reason that the commission voted against his recommendation, and so he concluded that "[i]ntelligent management of the deer herd depends, in the last analysis, on public understanding of the deer problem" (Leopold 1946a, 4). The theme of the importance of ecological education would later appear in "The Land Ethic" (Leopold 1949).

Drawing from Leopold's reflections on the conflict over deer policy, the following subprinciples can be added to the previous set:

6. Take the long view of conservation problems, recognizing that interests that seem to be served by a particular action often end up being undermined in the long term.
7. Gather applicable scientific information from relevant scientific disciplines (note the plural) and take it into account when developing policy.
8. Engender public understanding of the relevant science and its impacts.

Subprinciple 6, taking the long view of conservation problems, goes hand in glove with Subprinciple 5 concerning promoting and protecting land health, inasmuch as it concerns the land's capacity for self-renewal and its ability to support a diversity of life over time. Subprinciple 7, taking into account relevant science, was also implicit in Leopold's reflections on Coon Valley, but he states it much more explicitly in his reflections on deer policy, perhaps because it was that much closer to his own areas of research and work. Concerning Subprinciple 8, Leopold also thought that ecological education should be more widespread and that it should take an approach that was broad-based and value-driven.[5] Presumably specific understanding of a given issue would be enhanced with improved ecological education more generally.

These subprinciples, it should be clear, are mostly about the process of doing policy rather than a set of prescriptions of what policy should be. My contention in this chapter is that these subprinciples are practical, reasonable, and comprehensive, allowing for application to any conservation domain: wilderness practices, restorations, dealing with invasive species and climate change, how to farm, how to live in cities, and more. They are not, of course, a guarantee of success, as Leopold knew all too well. These claims are discussed further below. I begin with an elaboration of the subprinciples that seem most in need of further explication.

BALANCING VALUES AND INTERESTS

Subprinciples 1 and 2 might seem to be asking the impossible. After all, the variety of interests and values are sure to conflict. For example, one frequent sort of conflict occurs when an invasive species threatens the existence of other species and the health of the land community more generally, as happens with cats, kudzu, or cheatgrass.[6] How should we "integrate" those different interests and val-

---

5. Regarding the need for a broad-based approach, Leopold wrote, "An understanding of ecology does not necessarily originate in courses bearing ecological labels; it is quite as likely to be labeled geography, botany, agronomy, history, or economics" (Leopold 1949, 224). Regarding the need for a value-driven approach, Leopold expressed dissatisfaction with conservation education that "defines no right or wrong, assigns no obligation, calls for no sacrifice, implies no change in the current philosophy of values" (1949, 207–8).

6. The term *invasive species* is somewhat controversial and has been defined in a multitude of ways. I use it to mean any species that, through its reproduction or behavior, causes extirpation of other species, threatening the health of the entire community, regardless of whether the species is native or nonnative.

ues? How can we get advocates for particular species and advocates for land communities to cooperate rather than compete? How do we find a harmonious, balanced system of land use in such cases?

The obvious answer is to seek out win-win situations, and clearly that is the approach that would best satisfy Subprinciples 1 and 2. Thus, we should engage the diverse values, interests, and expertises in coming up with possible solutions to try to find a win-win (or as close as possible to a win-win) and not settle for the most obvious, or easiest, or cheapest solution. But what do we do when, despite the best efforts of all of relevant experts, no win-win situation can be found? (Chaigneau and Brown 2016).

Early philosophical scholarship implied that according to the land ethic, individuals should be sacrificed to wholes (e.g., Regan 1983), but that interpretation has since been debunked (e.g., Nelson 1996; Marietta 1999; Callicott 1999; Meine 2022; see chapter 1). Indeed, this is not a defensible interpretation of the land ethic, remembering always that the land ethic is explicitly intended to *expand* rather than replace our previous human ethics. But even with the debunking of that interpretation, the conflicts remain.

For many philosophers, the answer lies in creating additional rules to adjudicate between the competing interests (e.g., Shrader-Frechette 1996; Callicott 2014). Perhaps those are reasonable ways of proceeding, but my approach in this book aims to elucidate Leopold's own views as clearly as I can. Based on Leopold's policy activities, some of which are described in the first two sections of this chapter, I don't think that he would endorse an approach that was that rigid and inflexible. I take Leopold at his word when he enumerates various values in play in any conservation decision, including aesthetic and recreational values as well as the intrinsic value of species and communities. These are real and important values, none of which should always be routinely shunted aside for the others.

Indeed, following Don Marietta (1999), I think it is a mistake to ignore relevant sources of value by arbitrary fiat simply to achieve one clear answer. That is, there are good reasons to think that particular entities genuinely have value based on characteristics such as their autonomy, their ability to feel pain, or even simply because they are alive, but communities likewise have characteristics such as land health that are worthy of protection, as I argued in chapter 4. Although the easiest route is to declare categorically the ways in which some values trump other values, that route unjustifiably sacrifices some values for others.

If we are not to have an algorithm for weighing values that applies to all cases, then the only other way forward seems to be to characterize a *process* for coming to a decision in a particular case. The "bull sessions" that Leopold esteemed are instructive. Implicit in such sessions is the idea that people with different interests/values and areas of expertise (this is where scientific findings, Subprinciple 7, come in) are sharing their different techniques and approaches (Subprinciple 3), listening to and learning from one another. No doubt disagreements arise, and attempts are made to hash those out. But importantly, there is, as Leopold said, a sense of *common purpose*. As he argued elsewhere, "Conservation calls for something which the technologies, individually and collectively, now lack. . . . They lack, firstly, a collective purpose: stabilization of land as a whole. Until the *technologies accept as their common purpose the health of the land as a whole*, 'coordination' is mere window-dressing, and each will continue in part to cancel the other. The acceptance of this common purpose does not call for the surrender of their separate purposes (soil, timber, game, etc.) except as these conflict with the common one" (Leopold 1942a, 202; emphasis added).

So, the coordination that Leopold called for (Subprinciple 2) requires a common commitment to land health, which is a commitment to the long view (Subprinciples 5 and 6). Leopold surely recognized that the land would be healthier if we refrained from many of our contemporary human activities altogether, but he does not call for us to "surrender" our human purposes; he does not call for us to sacrifice ourselves for the health of the whole, even while recognizing that some sacrifices must be made. Rather, he insists only that proponents of the separate human purposes should try to harmonize with each other and not "conflict" with the common purpose of land health—a common purpose that exists because all members of the land community are interdependent with one another. After all, any human activity that seriously undermines land health would ultimately entail the inability to engage in that human activity. Agricultural practices that deplete the soil undermine our ability to farm in the future (thus the need for the Coon Valley Erosion Project); permitting the deer herd to get too large now would cause it to crash in the future. Indeed, Leopold wrote, "preoccupation with mankind, as distinguished from the community of which man is a member, defeats its own ends" (quoted in Meine 2010, 482); as noted in chapter 5, "picking and choosing" any one species is self-defeating, and that goes for always picking and choosing humans (i.e., anthropocentrism) as well. Thus, the process of creating a policy that incorporates diverse interests and values must have land health and the long view as the purpose held in common by all, even while recognizing (as noted above) that harmony in land use is an ideal to strive for.

Such a process might seem impossible, but there are contemporary cases that seem to fit. For example, the Ashland Forest Resiliency Project involved bringing together people with a seemingly intractable conflict of interests: anti-logging environmental activists and Forest Service employees accustomed to using timber sales of the biggest trees to finance the infrastructure for firefighting.[7] Neither side trusted the other: activists suspected that loggers were using forest fires as an excuse to log, and loggers suspected that activists would never agree to any tree cutting. Linda Duffy, the district ranger in charge in Ashland, Oregon, brought the two sides together with "a protracted series of community meetings, patience-stretching failures of communication, multiple compromises, and finally the crystallization of trust and understanding between groups that long thought they would never be able to work together" (Johnson 2021). They had a common purpose—wanting to protect the forest from fire in a time of climate change, drought, beetle infestation, crowded forests, and thick undergrowth. Moreover, Duffy thought that the activists "could offer fresh ideas, science, and resources that the Forest Service could not access on its own" (Johnson 2021). In the end, the activists agreed to a plan of action that involved selective cutting and maintenance burning, once they were convinced that the cutting was for protecting the forest for fire and not for revenue-producing logging. Fire experts from local Indigenous tribes played a key role in convincing the activists that some human intervention was necessary.

Duffy's actions in bringing together groups with diverse inter-

---

7. The following discussion of the Ashland Forest Resiliency Project relies on Johnson (2021).

ests, values, and expertise (activists, loggers, fire experts from Indigenous tribes) and encouraging them to listen to each other and focus on their common purpose—the health of the Ashland forest—exemplify a Leopoldian approach to policy as I have outlined it here. The Forest Service gave up on the idea that the project needed to pay for itself; activists gave up on the idea that all cutting was bad. Time will tell if the project will be successful. Leopold's approach cannot, of course, guarantee success. No approach to policy can guarantee that. But it was successful in another way. As one Ashland resident involved in the project put it, "Before community conversation, you had lawsuits.... You had no management happening. You had complete polarization. Even spending a couple of years in meetings and planning is a short-term investment for a long-term yield" (Johnson 2021). So, although adhering to the Principle of Integration of Land Uses might seem slow and is surely frustrating at times, it can result in an agreement that will stick. (It took about five years to come to the first stages of the plan, which has been evolving and expanding for more than fifteen years since then). One hopes that relying on diverse expertise will increase the chances of success for the plan, where success would be the long-term health of the Ashland forest, even in the face of ongoing burn threats and climate change.

As for the possibility of projects that simultaneously satisfy the diverse interests of humans, nonhumans, and the land community alike, the Yolo Bypass in Northern California's Sacramento Valley, close to the University of California, Davis, provides an example. The Yolo Bypass is an engineered floodplain on the same location as the historical, natural floodplain of the Sacramento River. Part of a network of weirs and bypasses, it is intended to "mimic the Sacramento River's natural floodplain functions" (Sommer et al. 2001, 7). It is typically flooded during the winter months (the rainy season in California). The Yolo Bypass serves a variety of functions: it has provided flood control that has "saved valley communities numerous times" (9); it has allowed for seasonal agriculture in the late spring and summer, with crops such as sugar beets, rice, safflower, and corn; it includes large areas of wetlands that are managed to provide habitat for migratory waterfowl and also provide habitat for various species of shorebirds, raptors, songbirds, and mammals, including threatened species; it is used for recreation and education (bird-watching, hiking, guided tours); and it provides key aquatic habitat for forty-two fish species, including fifteen native fish species, some of which are threatened or endangered (Sommer et al. 2001). Recent studies have focused on whether winter's flooded rice fields can serve as a rearing area for juvenile salmon, and the results so far are promising (Katz et al. 2017).

Of course, the Yolo Bypass project isn't "perfect." Proponents acknowledge that improvements could be made to its design and that the approach would not work in all regions (Sommer et al. 2001), although others maintain that "the potential of managing a working agricultural landscape for the combined benefits to fisheries, farming, flood protection, and native fish and wildlife species . . . should have broad applicability for the management of floodplains throughout California and beyond" (Katz et al. 2017, 13). Moreover, modifications are ongoing; recently, ground was broken on the Yolo Bypass Salmonid Habitat Restoration and Fish Passage (Big Notch) Project to "improve fish passage and increase floodplain fisheries rearing habitat in Yolo Bypass and the lower Sacramento River basin" via "a new Fremont Weir headworks structure, an outlet channel, and downstream channel improvements" (California Department of Water Resources). Once again, the commitment to the common purpose of land health, which is beneficial to all in the long run, is essential.

Note that the commitment to the common purpose of land health and the commitment to harmonize with the other interests mean that this process, *if properly followed*, would entail that destructive interests like those of the fossil fuel industry would not get to hold sway. They would somehow have to show how they could adhere to those commitments, perhaps motivated by seeing how it would be in their best long-term interests to do so. Moneyed interests, of course, have an outsized role in our actual policy decisions and often fail to "cooperate" with other interests; sadly, the Principle of Integration of Land Uses or any other account of policy cannot prevent that, necessitating political pressure and activism that is beyond the scope of this book to discuss.

Cases could be multiplied many times, of course, but I hope the ones I've given suggest how interests and values can be balanced. To be clear, my claim is not that these were inspired by Leopold, although they may have been. My point is to show that the Leopoldian approach to policy can be used, and used successfully, in the real world, and the cases illustrate how. In the next section I turn to another challenge for the Principle of Integration of Land Uses.

INCORPORATING LAND HEALTH INTO POLICY

The Ashland Forest Resiliency Project shows how people with multiple interests can come together and agree even in the face of initial disagreement; the Yolo Bypass shows how a project can fulfill many diverse interests, both human and nonhuman, simultaneously. But perhaps what is needed at this point is a clear example of how land health (Subprinciple 5) can be the result of such a process.

I think many contemporary organic farms implement this Leopoldian vision. Full Belly Farm located near Guinda, California, is one such example.[8] Full Belly Farm was founded in 1985 and consists of 250 acres; it is certified organic by California Certified Organic Farmers (SAGE [Sustainable Agriculture Education] and EPS [Economic & Planning Systems] 2007). According to one description of Full Belly Farm,

> The farm raises more than 80 different crops including vegetables, herbs, nuts, flowers, fruits, and grains and also raises chickens and sheep. The farm landscape is a diverse patchwork of annual crops, pastures and perennial orchards, hedgerows, and riparian areas managed as habitat for beneficial insects, native pollinators, and wildlife.
>
> The ecological diversity at Full Belly Farm was intentionally designed to foster sustainability on all levels. Production goals include healthy soil, a stable, fairly compensated work-force, an engaging workplace that renews and inspires everyone working on the farm, and happy customers. The productivity of this agroecosystem is based on the use of cover crops and the integration of sheep and poultry to capture and cycle crop nutrients and water, maintain soil health, and prevent losses from pests and disease. Virtually all of the production on the farm is irrigated, mostly

---

8. Full disclosure: I have been a Community Supported Agriculture (CSA) customer of Full Belly Farm since 2009 and was treated to a site visit in January 2020, where I was able to see and learn about many of these elements firsthand. I have no other connection to Full Belly Farm.

with water from Cache Creek, which runs along one side of the property. The farm sells to a diverse mix of direct markets in the San Francisco Bay area that includes restaurants, grocers, farmers' markets, and a 1500-member CSA. Full Belly also supports a number of educational and outreach programs to help create awareness of the importance of farms to all communities. (Lengnick et al. 2015, 580–81)

Clearly, Full Belly Farm supports a wide variety of plants, nonhuman animals, and humans in diverse ways, simultaneously satisfying large number of diverse values. That these practices also support land health is suggested by the emphasis on promoting biodiversity and soil fertility, the two main causes of land health that Leopold identified (see chapter 4). Indeed, some of these practices seem to fulfill Leopold's plea "that the wild life cover, at least on waste corners and fencerows, now become an expression of localized scientific reasoning and the owner's personal taste, rather than a badge of compliance to social regimentation" and his suggestion "that the slick and clean countryside is neither more beautiful, nor—in the long run—more useful than that which retains at least some remnants of non-domesticated plant and animal life" (Leopold 1938a, 78; emphasis added). Thus, diversity, in an agricultural context, consists of "a food chain aimed to harmonize the wild and the tame in *the joint interest of stability, productivity and beauty*" rather than a "food chain aimed solely at economic profit" (Leopold 1941d, 462; emphasis added), recalling that Leopold often used the terms *stability* and *land health* interchangeably (see chapter 4). In this way, Full Belly Farm goes beyond being just an organic farm to being one that actively promotes biodiversity (see, e.g., Tscharntke et al. 2021 for discussion of this farming approach).[9]

Furthermore, these practices have been successful in promoting land health. In 2014, after five years of prolonged drought, "many vegetable growers in the region had to take land out of production, invest in new wells, or transition to higher value tree crops in an attempt to stay in business" while "Full Belly Farm remained productive and profitable" (Lengnick et al. 2015, 578). Why? "High soil

---

9. Leopold coined the phrase "biotic farming" to refer to farming with the "largest possible diversity of flora and fauna" (1939a, 730).

quality, biodiversity, and profitable direct markets appear to be key to the farm's resilience to the continuing drought and more frequent winter flooding that plagues the region" (578).[10] Another illustration of the way in which Full Belly Farm promotes land health is in its use and support of pollinators. According to a Xerces Society publication, "[c]over crops, hedgerows, untended corners, and a diverse, organic cropping system are all tools used by Full Belly to benefit crop pollinators and the bottom line. . . . The result of the location and practices is a farm that does not need to rent or manage a single honey bee hive" (Vaughan et al. 2007, 14). This epitomizes the capacity of land for self-renewal (i.e., land health). One might think that the main purpose of a farm is to produce food for profit, and that is not a mistaken view, but Full Belly Farm shows that to be truly sustainable over the long run, a farm must also work toward the "common purpose" of maintaining the health of the land (especially, on a farm, its soil fertility).[11]

Given all of this, it should perhaps not come as a surprise that in 2014, Full Belly Farm was awarded the California Leopold Conservation Award with the following affirmation: "When it comes to farming in ways that promote the long-term health of California's land, water, wildlife and food economy, there's no better example than Full Belly Farm" (Sustainable Conservation 2014). But perhaps the reader might not consider Full Belly to be an example of "policy," given that it is a privately owned farm. Here it is important to recognize that Leopold came to believe, as he emphasized in "The Land Ethic" and elsewhere (including his discussion of the Coon Valley Erosion Project), that conservation cannot simply be the province of governments on public lands but must also involve the practices and values of individuals on private lands. A full account of conservation policy, Leopold implied, must include both.

---

10. Given that climate change is expected to continue to worsen drought in California, Full Belly's managers continue to tweak their practices (e.g., which cover crops, what sort of irrigation) in order to maintain the health of the farm (Lengnick et al. 2015); there is no magic set of land health-promoting practices.

11. I am setting aside the important but complicated empirical question—which it is beyond the scope of this book to address—of how a farm like this would "scale up," or even if "scaling up" is the right answer. No doubt many changes would be desirable for food production, including changes in consumer habits.

Even on private land, the Principle of Integration of Land Uses still applies, meaning that a multitude of values and approaches must be taken into account with the common purpose of promoting and maintaining land health.

A very different sort of example that has exercised conservation biologists and environmental ethicists for decades is that of feral pigs (Sus scrofa) in Hawaii, because it brings together issues of a human-introduced and invasive predator; a highly biodiverse, unique, and sensitive biotic community; Indigenous rights; and even animal rights. Ongoing and recent work might seem to suggest that eradicating feral pigs would improve land health in Hawaii. For example, in a study on Kaua'I, researchers conclude that although feral pig depredations of seabirds are not as common as those of cats and rats, "when they did occur they were devastating" because they "typically involved the destruction of the entire burrow, which the pigs excavated to access whatever was inside: . . . adults, chicks, or eggs" (Raine at al. 2020, 432). Moreover, they found that predator control methods were effective. Another study found that, contrary to initial expectations, removing feral pigs increased the diversity of soil bacterial communities, with diversity scores positively correlated with time since removal (Wehr et al. 2019). Studies such as these (and there are many) might *seem* to suggest that a Leopoldian policy based on the land ethic would advocate for the removal of the feral pigs in some fashion given the harmful effect of the feral pigs on land health in Hawaii.

Yet that is not the most Leopoldian option. According to Luat-Hū'eu et al., the "majority of studies [of feral pigs in Hawaii] primarily focused on the ecological impacts of feral pigs without accounting for sociocultural values" (Luat-Hū'eu et al. 2021, 443). The study recommends "that government agencies work in greater collaboration with Indigenous peoples and local communities, particularly hunters, using multi-objective approaches to manage culturally-valued nonnative species like feral pigs, alongside other culturally-valued native species, to reduce conflicts among stakeholders and support the perpetuation of all cultural practices" (448), given that Indigenous peoples have been hunting the feral pigs since the mid-nineteenth century. Indigenous peoples *also* value "native biodiversity and the integrity of native-dominated landscape" (448). Thus, in line with the arguments of this chapter, I suggest that a Leopoldian conservation policy would support the call that Luat-Hūʻeu and colleagues make for a scientifically informed, multi-valued, multi-interest dialogue to determine policies concerning feral pigs in Hawaii, with the understanding that the common purpose—but not the sole purpose—is the land health of the Hawaiian land communities. Such policies could not allow the feral pigs free range, but they could allow for hunting by Indigenous peoples, perhaps in specified locations or in specified ways. Indeed, reflecting on Leopold's work on the Wisconsin Conservation Commission to determine a deer-hunting policy that would simultaneously maintain land health, I believe he would have welcomed the development of policies that included, or at least took into consideration, hunting interests.

Of course, some animal rights proponents would not support such policies if they still ended up including the killing of pigs. Still, I think there could be a role for an animal rights perspective in making control of the feral pigs as painless as possible. It is also worth considering that, just as it is not in the long-term interests of deer populations to grow very large, it may not be in the long-term interests of feral pig populations to grow unchecked, because it could lead to suffering and death of the pigs—and other animals—in the long run. This is just to reiterate Leopold's point that obligations to promote land health cannot and should not be overlooked.

Here I mention other sorts of cases briefly.[12] As discussed in chapter 4, the eradication of wolves from Yellowstone National Park and elsewhere is widely seen as having contributed to land sickness, and the restoration of wolves to Yellowstone is widely seen as having successfully restored land health, since it has benefited many other species (via controlling the size of elk populations and thus their impacts), reduced erosion (by changing elk behavior), and improved water flow. Chapter 4 also mentions that the land ethic would emphasize approaches to addressing climate change that *also* promote land health. Not all proposals for mitigating climate change would do this, but many would. Shin et al. assert that "conservation actions

---

12. As with the last section's cases, I am not claiming that these approaches were directly inspired by Leopold; they are meant to be illustrative.

that halt, slow or reverse biodiversity loss can simultaneously slow anthropogenic mediated climate change significantly" (Shin et al. 2022, 2847). They identify fourteen out of the twenty-one action targets of the draft post-2020 global biodiversity framework of the Convention on Biological Diversity as having co-benefits for addressing biodiversity loss and climate change mitigation. They suggest that "[a]voiding deforestation and restoring ecosystems (especially high-carbon ecosystems such as forests, mangroves or seagrass meadows) are among the conservation actions having the largest potential for mitigating climate change" while emphasizing that local needs and socioeconomic contexts must also be taken into account (Shin et al. 2022, 2848). This multi-valued approach to climate change that prioritizes land health both locally and globally would be a policy priority for a Leopoldian approach.

More generally, incorporating land health into policy means looking for the causes of land health (biodiversity and soil fertility) or lack thereof, as well as the common symptoms of land sickness, such as abnormal erosion and abnormal flooding; it also means looking for the presence or absence of long food chains forming a tangled web of interdependencies.[13] Here, though, it is important to recall that these were Leopold's hypotheses, not claims that he thought he had sufficient evidence to defend with certainty. Thus, if it were to turn out that there were other ways to promote land health, these would be consistent with a Leopoldian approach: the Principle of Integration of Land Uses.

OTHER POTENTIAL ISSUES OF CONCERN

One question that might arise is the extent to which humans should take an active role in maintaining "natural areas," including those that we have restored, or rehabilitated, or "greened." I think there is no simple answer to this question—and again this should be a matter taken up on a case-by-case basis by those with a variety of relevant interests—but here are some considerations. One is a reminder that Leopold clearly saw humans as interdependent with

---

13. See chapter 4 for a discussion of what Leopold saw as the causes of land health/sickness and the symptoms of land sickness.

other members of the land community and thus as parts of the land community (see chapters 1 and 2 for further discussion). So there is nothing in principle wrong with humans taking an active role. Indeed, it might be argued that since our negative impacts are ongoing—climate change in particular—that positive interventions might likewise need to be ongoing. Moreover, even though *Homo sapiens* is a "plain member and citizen" of the land community (Leopold 1949, 204), all of our fellow members and citizens are contributing via their distinctive ecological interactions and interdependencies. We ought to do likewise by contributing in the distinctive ways that we can, as beings that have some capacity to evaluate situations, learn, make choices about the best course of action, and implement solutions as best as we can.[14]

Nonetheless, as I have emphasized in previous chapters, Leopold thought that we are often ignorant of many interdependencies and thus ought to proceed cautiously, making changes slowly and moderately and in line with how land communities developed to behave over millennia, to the extent we can. He also worried about an "increasing dependence on artificial replenishment from hatcheries and propagating plants, and on artificial control of 'undesirable' species" because "artificial replenishment and control are always costly and often ineffective"; instead, we should modify land use so as to provide habitat for each species (Leopold 1941c, 195). Providing the required habitats, he thought, would "reduce the need for artificial interference" by allowing for a restoration of the interdependencies that help keep species flourishing at reasonable numbers (Leopold 1941c, 195).

On a related note, it should be clear from this chapter that Leopoldian policy goes far beyond setting aside wilderness. That being said, Leopold clearly did think that *some* wilderness should be set aside, but we need to have a careful understanding of what he meant by wilderness. He recognized that "[m]any of the diverse wilder-

---

14. Arguably, Leopold's remarks from "On a Monument to a Pigeon" are exactly this, an exhortation to use our distinctive abilities as humans: "To love what *was* is a new thing under the sun, unknown to most people and to all pigeons. To see America as history, to conceive of destiny as a becoming, to smell a hickory tree through the still lapse of ages—all these things are possible for us, and to achieve them takes only the free sky, and the will to ply our wings" (Leopold 1949, 112).

nesses out of which we have hammered America are already gone" (1949, 121) but thought that there were remnants of varying sizes and degrees of wildness, and that a "representative series of these areas can and should be kept" (122) for recreation, for science, and for wildlife (see chapter 1). But setting aside wilderness is not the central or sole focus of the land ethic. As Meine argues, "Leopold as a conservation thinker, scientist, advocate, and practitioner never focused exclusively on wildland protection" (Meine 2022, 171; see also Meine 2017).

Another sort of concern that might arise is whether the policy subprinciples I have outlined and defended here are substantive enough to provide guidance; they might seem vague, or weak, or too open-ended. One way to respond to this concern would be to point out that "[a] land ethic, as interpreted and extended since Leopold's time, has inspired innumerable community-based conservation efforts and locally driven movements, on behalf of everything from food sovereignty to watershed rehabilitation to urban land restoration" (Meine 2022). In other words, it is clear that the land ethic can provide powerful guidance, because it has already been so successful at implementing policy.

An additional way to respond to a concern over the worth of the eight subprinciples is to contrast these principles with the guidance that would follow from more traditional accounts of environmental ethics—for example, one that always prioritized humans, or always prioritized sentient animals, or one that always prioritized individual living beings, or one that always prioritized the ecosystem as a whole.[15] That land-ethic-inspired principles do not take any of these positions shows that they in fact represent a definitive stand—a definitive stand that allows for flexibility to respond to different situations (different facts on the ground and different contexts) and different values in play. They embody a definitive stand that urges people with different interests to work together to try to find creative solutions to address the different values in play, not just the easiest or most obvious solution. A definitive stand that, although it may require more time to reach an agreement, is more

---

15. Again, Leopold would find any of these approaches ethically insufficient and/or ultimately self-defeating.

likely to produce a solution that will stand the test of time both in the eyes of those most directly affected (which is necessary for success) as well as for the long run of the various interests in play (i.e., the common purpose, the health of the land).

The Leopoldian policy subprinciples I have outlined here also differ from the "systematic conservation planning" approach of Watson et al. (2011) and similar approaches. The overall Leopoldian approach is comparable in gathering diverse "stakeholder" input, but Subprinciples 5 and 6 insist that land health and the long view (the common interests) not be subverted to individual interests. To the extent that the systematic conservation planning approach might allow a subversion like that (and it is not fully clear to me how much it might), it differs from the Leopoldian approach defended in this chapter. Also different is the "global public goods" approach described by Brando et al. (2019), which typically adopts the anthropocentric approach that Leopold described as self-defeating (as any overriding focus on individual species is). I mention these mainly to further illustrate that Leopold's approach to policy stakes out definitive ground that differs from some contemporary approaches while being in sync with the others described in previous sections.

Finally, any substantive approach to policy ought to be able to identify policy processes that have gone wrong. A recent decision in Foster City, California, provides an example. As described in the *San Francisco Chronicle*, the city has been experiencing a "problem" with Canadian geese (guano contamination); they settled on a "solution" of a lethal spinal dislocation rather than trying a multitude of nonlethal methods (including eliminating the attractive turf grass), saying that "the problem had gone on too long and that the time for 'talk and discussions' was over" (Vainshtein 2022). Here was a case where diverse interests and values were expressed by citizens, where alternatives were proposed that might satisfy many of those interests and values, yet city officials went in a different direction in the name of expediency. This is not a Leopoldian approach to policy, and it is one that arguably caused needless suffering to the geese and to the citizens who were concerned about them; moreover, it is probably not a long-term solution, since more geese are likely to migrate to the area.

Of course, as I have emphasized already, success is not guaranteed with a Leopoldian approach to policy, either in the process or the outcome; policy is hard. Leopold's own experiences show that. But no set of policy guidelines can guarantee a good process or a good outcome. My claim, then, is that the eight Leopold-inspired subprinciples outlined in this chapter provide for a defensible and inclusive process—the Principle of Integration of Land Uses—that can guide us through the variety of challenges that we face as a society today. Moreover, my claim is that these principles can guide us in a scientifically and ethically justified way, inasmuch as they are grounded in the land ethic.

CONCLUSION

Some of the policy implications of the land ethic follow directly from what has already been argued earlier in the book: namely, that a land ethic implies that we ought to seek to preserve land health, which implies that we sustain species populations ("biodiversity") and their consequent ecological interactions and interdependencies as well as matter/energy flows, remembering that abiotic components (particularly the soil) are part of land health as well. These practices apply not just to "wilderness" areas, but to all human practices, including farming, forestry, restoration, even how we live in cities. But more than protecting land communities—recall that the land ethic is an extension of our existing ethics—the land ethic calls on us to protect individual human and nonhuman organisms as well. (That is, the land ethic is a pluralist ethic, recognizing multiple entities of value). Of course, conflicts arise between competing values, and we often cannot fully respect the rights of all. During his lifetime, Leopold dealt with such conflicts head on, by engaging and negotiating with people who had diverse interests and values with regard to the issue at hand. From his experiences, I derive the eight subprinciples of the Principle of Integration of Land Uses articulated above, elaborating and defending them. They take into account all the entities of value in concert with the concrete facts of a situation rather than, say, arguing for a particular algorithm that would determine which entities should take precedence. They provide a reasonable, flexible, and comprehensive approach to policy that always keeps in mind the long view. They encourage public education and public participation.

The land ethic is grounded in well-founded and scientifically supported core concepts—interdependence (from which it all stems), land community, and land health—and it is supported by a simple yet powerful and persuasive argument that yields a practical and practicable approach to policy. It is thus conceptually, ethically, scientifically, and pragmatically supported. It is a land ethic for our time.

References

Andresen, Ellen, Víctor Arroyo-Rodríguez, and Federico Escobar. 2018. "Tropical Biodiversity: The Importance Of Biotic Interactions For Its Origin, Maintenance, Function, and Conservation." In *Ecological Networks in the Tropics: An Integrative Overview of Species Interactions from Some of the Most Species-Rich Habitats on Earth*, edited by W. Dátillo and V. Rico-Gray, 1–13. Cham, Switzerland: Springer.

Anstett, Marie-Charlotte, Martine Hossaert-McKey, and Doyle McKey. 1997. "Modeling the Persistence of Small Populations of Strongly Interdependent Species: Figs and Fig Wasps." *Conservation Biology* 11 (1): 204–13.

Bell, Daniel. 2020. "Communitarianism." In *The Stanford Encyclopedia of Philosophy* (Fall 2022 Edition), edited by Edward N. Zalta.

Berkes, Fikret, Nancy C. Doubleday, and Graeme S. Cumming. 2012. "Aldo Leopold's Land Health from a Resilience Point of View: Self-Renewal Capacity of Social–Ecological Systems." *EcoHealth* 9 (3): 278–87.

Beschta, Robert L., Luke E. Painter, and William J. Ripple. 2018. "Trophic Cascades at Multiple Spatial Scales Shape Recovery of Young Aspen in Yellowstone." *Forest Ecology and Management* 413:62–69.

Beschta, Robert L., and William J. Ripple. 2012. "Berry-Producing Shrub Characteristics Following Wolf Reintroduction in Yellowstone National Park." *Forest Ecology and Management* 276:132–38.

Binkley, Dan, Margaret M. Moore, William H. Romme, and Peter M. Brown. 2006. "Was Aldo Leopold Right about the Kaibab Deer Herd?" *Ecosystems* 9:227–41.

Boorse, Christopher. 1977. "Health as a Theoretical Concept." *Philosophy of science* 44 (4): 542–73.

Bouyer, Jeremy, Sophie Ravel, Jean-Pierre Dujardin, Thierry De Meeüs, Laurence Vial, Sophie Thévenon, Laure Guerrini, Issa Sidibé, and Philippe Solano. 2007. "Population structuring of *Glossina palpalis gambiensis* (Diptera: Glossinidae) According to Landscape Fragmentation in the Mouhoun River, Burkina Faso." *Journal of Medical Entomology* 44 (5): 788–95.

Brando, Nicolás, Christiaan Boonen, Samuel Cogolati, Rutger Hagen, Nils Vanstappen, and Jan Wouters. 2019. "Governing as Commons or as Global Public Goods: Two Tales of Power." *International Journal of the Commons* 13 (1).

Brooker, Rob W., and Terry V. Callaghan. 1998. "The Balance between Positive and Negative Plant Interactions and Its Relationship to Environmental Gradients: A Model." *Oikos* 81 (1): 196–207.

Cahill, Abigail E., Matthew E. Aiello-Lammens, M. Caitlin Fisher-Reid, Xia Hua, Caitlin J. Karanewsky, Hae Yeong Ryu, Gena C. Sbeglia, Fabrizio Spagnolo, John B. Waldron, Omar Warsi, and John J. Wiens. 2013. "How Does Climate Change Cause Extinction?" *Proceedings of the Royal Society of London B: Biological Sciences* 280 (1750); https://doi.org/10.1098/rspb.2012.1890.

California Department of Water Resources,. 2022. "Riverine Habitat Restoration Projects." https://cwc.ca.gov/Home/Programs/Integrated-Science-and-Engineering/Restoration-Mitigation-Compliance/Yolo-Bypass-Projects.

Callicott, J. Baird. 1987. "The Conceptual Foundations of the Land Ethic." In *Companion to A Sand County Almanac*, edited by J. Baird Callicott, 186–217. Madison, WI: University of Wisconsin Press.

———. 1989. "Hume's Is/Ought Dichotomy and the Relation of Ecology to Leopold's Land Ethic." In *In Defense of the Land Ethic: Essays in Environmental Philosophy*, edited by J. Baird Callicott. Albany: State University of New York Press.

———. 1995. "The Value of Ecosystem Health." *Environmental Values* 4 (4): 345–61.

———. 1997. "Fallacious Fallacies and Nonsolutions: Comment on Kristin Shrader-Frechette's 'Ecological Risk Assessment and Ecosystem Health: Fallacies and Solutions.'" *Ecosystem Health* 3 (3): 133–35.

———. 1999. *Beyond the Land Ethic: More Essays in Environmental Philosophy*. Albany: State University of New York Press.

———. 2014. *Thinking like a Planet: The Land Ethic and the Earth Ethic*. Oxford: Oxford University Press.

Callicott, J. Baird, and Karen Mumford. 1997. "Ecological Sustainability as a Conservation Concept." *Conservation Biology* 11 (1): 32–40.

Callicott, J. Baird, and Eric T. Freyfogle, eds. 1999. *Aldo Leopold: For the Health of the Land: Previously Unpublished Essays and Other Writings*. Washington, DC: Island Press.

Chaigneau, Tomas, and Katrina Brown. 2016. "Challenging the Win-Win Discourse on Conservation and Development: Analyzing Support for Marine Protected Areas." *Ecology and Society* 21 (1): 36.

Chambers, Jeanne C., Craig R. Allen, and Samuel A. Cushman. 2019. "Operationalizing Ecological Resilience Concepts for Managing Species and Ecosystems at Risk." *Frontiers in Ecology and Evolution* 7:241.

Chapin III, F. Stuart, Pamela A. Matson, and Peter Vitousek. 2011. *Principles of Terrestrial Ecosystem Ecology*. New York: Springer Science & Business Media.

Clements, Frederic E. 1916. *Plant Succession: An Analysis of the Development of Vegetation*. Washington, DC: Carnegie Institution of Washington.

Commoner, Barry. 1971. *The Closing Circle: Nature, Man, and Technology*. New York: Knopf.

Costanza, Robert, and Michael Mageau. 1999. "What Is a Healthy Ecosystem?" *Aquatic Ecology* 33 (1): 105–15.

Craige, Betty Jean. 2002. *Eugene Odum: Ecosystem Ecologist and Environmentalist*. Athens: University of Georgia Press.

Currie, William S. 2011. "Units of Nature or Processes Across Scales? The Ecosystem Concept at Age 75." *New Phytologist* 190 (1): 21–34.

deLaplante, Kevin, and Valentin Picasso. 2011. "The Biodiversity-Ecosystem Function Debate in Ecology." In *Handbook of the Philosophy of Science: Philosophy of Ecology*, edited by Kevin deLaplante, Bryson Brown, and Kent A. Peacock, 169–200. North Holland: Elsevier.

Diamond, Jared. 1986. "Overview: Laboratory Experiments, Field Experiments, and Natural Experiments." In *Community Ecology*, edited by Jared Diamond and Ted J. Case, 3–22. New York: Harper & Row.

Donovan, Geoffrey H., David T. Butry, Yvonne L. Michael, Jeffrey P. Prestemon, Andrew M. Liebhold, Demetrios Gatziolis, and Megan Y. Mao. 2013. "The Relationship between Trees and Human Health: Evidence from the Spread of the Emerald Ash Borer." *American Journal of Preventive Medicine* 44 (2): 139–45.

Dussault, Antoine C. 2018. "Welfare, Health, and the Moral Considerability of Nonsentient Biological Entities." *Les Ateliers de l'Éthique / The Ethics Forum* 13 (1): 184–209.

———. 2021. "Trois Faux Dilemmes dans le Débat sur la Santé Écosystémique." In *Protéger l'Environnement: De la Science à l'Action*, edited by Ely Mermans and Antoine C. Dussault, 173–209. Paris: Éditions Matériologiques.

Eliot, Christopher H. 2011. "The Legend of Order and Chaos: Communities and Early Community Ecology." In *Handbook of the Philosophy of Science: Philosophy of Ecology*, edited by Kevin deLaplante, Bryson Brown, and Kent A. Peacock, 49–108. North Holland: Elsevier.

———. 2013. "Ecological Objects for Environmental Ethics." In *Linking Ecology and Ethics for a Changing World: Values, Philosophy, and Action*, edited by Ricardo Rozzi, S. T. A. Pickett, Clare Palmer, Juan J. Armesto, and J. Baird Callicott, 219–29. Dordrecht: Springer.

Elton, Charles. 1927. *Animal Ecology*. New York: Macmillan.

Fimrite, Peter. 2018. "Monterey Bay Aquarium's Program to Save Sea Otters Revives Their Habitat." *San Francisco Chronicle*, April 21, 2018.

Flader, Susan L. 1994. *Thinking like a Mountain: Aldo Leopold and the Evolution of an Ecological Attitude toward Deer, Wolves, and Forests*. Madison: University of Wisconsin Press.

———. 2011. "Aldo Leopold and the Land Ethic: An Argument for Sustaining Soils." In *Sustaining Soil Productivity in Response to Global Climate Change: Sci-*

ence, Policy, and Ethics, edited by Thomas J. Sauer, John Norman, and Mannava V. K. Sivakumar, 43–65. Hoboken, NJ: John Wiley & Sons.

Flader, Susan L., and J. Baird Callicott, eds. 1991. *The River of the Mother of God, and Other Essays by Aldo Leopold*. Madison: University of Wisconsin Press.

Fleming, P. J. S. 2019. "They Might Be Right, but Give No Strong Evidence That 'Trophic Cascades Shape Recovery of Young Aspen in Yellowstone National Park': A Fundamental Critique of Methods." *Forest Ecology and Management* 454:117283.

Frank, Kenneth T., Brian Petrie, Jae S. Choi, and William C. Leggett. 2005. "Trophic Cascades in a Formerly Cod-Dominated Ecosystem." *Science* 308 (5728): 1621–23.

Freyfogle, Eric T. 2008. "Land Ethic." In *Encyclopedia of Environmental Ethics and Philosophy*, edited by J. Baird Callicott and Robert Frodeman. Farmington Hills, Michigan: Macmillan Reference USA.

Fukami, Tadashi. 2015. "Historical Contingency in Community Assembly: Integrating Niches, Species Pools, and Priority Effects." *Annual Review of Ecology, Evolution, and Systematics* 46:1–23.

Ghiselin, Michael T. 1974. "A Radical Solution to the Species Problem." *Systematic Zoology* 23 (4): 536–44.

———. 1997. *Metaphysics and the Origin of Species*. Albany: State University of New York Press.

Ginn, Franklin, Uli Beisel, and Maan Barua. 2014. "Flourishing with Awkward Creatures: Togetherness, Vulnerability, Killing." *Environmental Humanities* 4 (1): 113–23. https://doi.org/10.1215/22011919-3614953.

Gleason, Henry Allan. 1917. "The Structure and Development of the Plant Association." *Bulletin of the Torrey Botanical Club* 44 (10): 463–81.

Goodpaster, Kenneth E. 1978. "On Being Morally Considerable." *Journal of Philosophy* 75 (6): 308–25.

———. 1979. "From Egoism to Environmentalism." In *Ethics and Problems of the 21st Century*, edited by Kenneth M. Sayre and Kenneth E. Goodpaster, 21–35. South Bend, IN: University of Notre Dame Press,.

Gounand, Isabelle, Chelsea J. Little, Eric Harvey, and Florian Altermatt. 2018. "Cross-Ecosystem Carbon Flows Connecting Ecosystems Worldwide." *Nature Communications* 9 (1): 1–8.

Gray, Joe, Ian Whyte, and Patrick Curry. 2017. "Ecocentrism: What It Means and What It Implies." *Ecological Citizen* 1 (2): epub-010.

Grosholz, Edwin, Gail Ashton, Marko Bradley, Chris Brown, Lina Ceballos-Osuna, Andrew Chang, Catherine de Rivera, Julie Gonzalez, Marcella Heineke, Michelle Marraffini, Linda McCann, Erica Pollard, Ian Pritchard, Gregory Ruiz, Brian Turner, and Carolyn Tepolt. 2021. "Stage-Specific Overcompensation, the Hydra Effect, and the Failure to Eradicate an Invasive Predator." *Proceedings of the National Academy of Sciences* 118 (12): e2003955118. https://doi.org/doi:10.1073/pnas.2003955118.

Guha, Ramachandra. 1989. "Radical American Environmentalism and Wilderness Preservation: A Third World Critique." *Environmental Ethics* 11 (1): 71–83.

Harley, C. D. G., and M. D. Bertness. 1996. "Structural Interdependence: An Ecological Consequence of Morphological Responses to Crowding in Marsh Plants." *Functional Ecology* 10:654–61.

Hastings, Alan, and Louis Gross. 2012. *Encyclopedia of Theoretical Ecology*. Berkeley, CA: University of California Press.

Helms, Douglas. 1992. "Coon Valley, Wisconsin: A Conservation Success Story." In *Readings in the History of the Soil Conservation Service*, 51–53. Washington, DC: Soil Conservation Service.

Hettinger, Ned, and Bill Throop. 1999. "Refocusing Ecocentrism: De-emphasizing Stability and Defending Wildness." *Environmental Ethics* 21:3–21.

Hoek, Tim A., Kevin Axelrod, Tommaso Biancalani, Eugene A. Yurtsev, Jinghui Liu, and Jeff Gore. 2016. "Resource Availability Modulates the Cooperative and Competitive Nature of a Microbial Cross-Feeding Mutualism." *PLoS Biology* 14 (8): e1002540.

Holder, Philippa J., Ainsley Jones, Charles R. Tyler, and James E. Cresswell. 2018. "Fipronil Pesticide as a Suspect in Historical Mass Mortalities of Honey Bees." *Proceedings of the National Academy of Sciences* 115 (51): 13033–38.

Holling, Crawford S. 1973. "Resilience and Stability of Ecological Systems." *Annual Review of Ecology and Systematics* 4:1–23.

Holling, C. S., and Gary K. Meffe. 1996. "Command and Control and the Pathology of Natural Resource Management." *Conservation Biology* 10 (2): 328–37.

Hooper, David U., F. S. Chapin III, John J. Ewel, Andrew Hector, Pablo Inchausti, Sandra Lavorel, John Hartley Lawton, D. M. Lodge, Michel Loreau, and Shahid Naeem. 2005. "Effects of Biodiversity on Ecosystem Functioning: A Consensus of Current Knowledge." *Ecological Monographs* 75 (1): 3–35.

Hornbeck, Richard. 2012. "The Enduring Impact of the American Dust Bowl: Short-And Long-Run Adjustments to Environmental Catastrophe." *American Economic Review* 102 (4): 1477–1507.

Huber, Machteld, J. André Knottnerus, Lawrence Green, Henriëtte Van Der Horst, Alejandro R. Jadad, Daan Kromhout, Brian Leonard, Kate Lorig, Maria Isabel Loureiro, and Jos W. M. Van Der Meer. 2011. "How Should We Define Health?" *BMJ* 2011;343:d4163. https://doi.org/10.1136/bmj.d4163.

Hull, David L. 1976. "Are Species Really Individuals?" *Systematic Zoology* 25: 174–91.

———. 1978. "A Matter of Individuality." *Philosophy of Science* 45:335–60.

Jian, Jinshi, Xuan Du, and Ryan D. Stewart. 2020. "A Database for Global Soil Health Assessment." *Scientific Data* 7 (1): 1–8.

Johnson, Nathaneal. 2021. "How One Town Put Politics Aside to Save Itself from Fire." Grist. https://grist.org/extreme-weather/how-one-town-put-politics-aside-to-save-itself-from-fire-ashland-oregon/.

Kadoya, Taku, and Kevin S. McCann. 2015. "Weak Interactions and Instability Cascades." *Scientific Reports* 5 (1): 12652. https://doi.org/10.1038/srep12652.

Katz, Eric. 1996. *Nature as Subject: Human Obligation and Natural Community*. Washington, DC: Rowman & Littlefield.

Katz, Jacob V. E., Carson Jeffres, J. Louise Conrad, Ted R. Sommer, Joshua Martinez, Steve Brumbaugh, Nicholas Corline, and Peter B. Moyle. 2017. "Floodplain Farm Fields Provide Novel Rearing Habitat for Chinook Salmon." PLoS ONE 12 (6): e0177409. https://doi.org/10.1371/journal.pone.0177409.

Kimmerer, Robin. 2013. *Braiding Sweetgrass: Indigenous Wisdom, Scientific Knowledge and the Teachings of Plants*. Minneapolis, MN: Milkweed Editions.

King, Martin Luther, Jr. 1963. "Letter from the Birmingham Jail." In *Why We Can't Wait*, edited by Martin Luther King Jr., 77–100. Cambridge, MA: Signet.

Kingma, Elselijn. 2010. "Paracetamol, Poison, and Polio: Why Boorse's Account of Function Fails to Distinguish Health and Disease." British Journal for the Philosophy of Science 61 (2): 241–64.

Kitchell, James F., Daniel E. Schindler, Brian R. Herwig, David M. Post, Mark H. Olson, and Michael Oldham. 1999. "Nutrient Cycling at the Landscape Scale: The Role of Diel Foraging Migrations by Geese at the Bosque del Apache National Wildlife Refuge, New Mexico." Limnology and Oceanography 44 (3): 828–36.

Kloor, Keith. 2015. "The Battle for the Soul of Conservation Science." Issues in Science and Technology 31 (2).

Knight, Richard L. 1996. "Aldo Leopold, the Land Ethic, and Ecosystem Management." Journal of Wildlife Management 60 (3): 471–74.

Kone, Naferima, Jérémy Bouyer, Sophie Ravel, Marc J. B. Vreysen, Kouadjo T. Domagni, Sandrine Causse, Philippe Solano, and Thierry De Meeûs. 2011. "Contrasting Population Structures of Two Vectors of African Trypanosomoses in Burkina Faso: Consequences for Control." PLoS Neglected Tropical Diseases 5 (6): e1217.

Lackey, Robert T. 1996. "Pacific Salmon, Ecological Health and Public Policy." Ecosystem Health 2 (1): 61–68.

Laundré, John W., Lucina Hernández, and William J. Ripple. 2010. "The Landscape of Fear: Ecological Implications of Being Afraid." Open Ecology Journal 3:1–7.

Le Conte, Yves, and Maria Navajas. 2008. "Climate Change: Impact on Honey Bee Populations and Diseases." Revue Scientifique et Technique-Office International des Epizooties 27 (2): 499–510.

Leibold, Mathew A., M. Holyoak, N. Mouquet, P. Amarasekare, J. M. Chase, M. F. Hoopes, R. D. Holt, J. B. Shurin, R. Law, and D. Tilman. 2004. "The Metacommunity Concept: A Framework for Multi-Scale Community Ecology." Ecology Letters 7 (7): 601–13.

Lengnick, Laura, Michelle Miller, and Gerald G. Marten. 2015. "Metropolitan Foodsheds: A Resilient Response to the Climate Change Challenge?" Journal of Environmental Studies and Sciences 5 (4): 573–92.

Leopold, Aldo. 1923. "Some Fundamentals of Conservation in the Southwest." In Flader and Callicott 1991, 86–97.

———. 1934. "The Arboretum and the University." In Flader and Callicott 1991, 209–11.

———. 1935a. "Coon Valley: An Adventure in Cooperative Conservation." In Callicott and Freyfogle 1999, 47–54.

———. 1935b. "Land Pathology." In Flader and Callicott 1991, 212–17.

———. 1937. "Conservationist in Mexico." In Meine 2013a, 394–99.

———. 1938a. "Be Your Own Emperor." In Callicott and Freyfogle 1999, 70–81.

———. 1938b. "A Survey of Conservation." In Meine 2013a, 416–24.

———. 1939a. "A Biotic View of Land." *Journal of Forestry* 37 (9): 727–30.

———. 1939b. "The Farmer as a Conservationist." In Meine 2013a, 425–37.

———. 1941a. "Conservation of Natural Resources." *The Aldo Leopold Archives*, http://digital.library.wisc.edu/1711.dl/AldoLeopold, Lectures 1933–1947: 289–90.

———. 1941b. "Lakes in Relation to Terrestrial Life Patterns." In *A Symposium on Hydrobiology*, edited by James Needham et al., 17–22. Madison: University of Wisconsin Press.

———. 1941c. "Planning for Wildlife." In Callicott and Freyfogle 1999, 193–98.

———. 1941d. "The Round River: A Parable of Conservation." In Meine 2013a, 458–62.

———. 1942a. "Biotic Land Use." In Callicott and Freyfogle 1999, 198–207.

———. 1942b. "Land-Use and Democracy." In Flader and Callicott 1991, 295–300.

———. 1942c. "The Role of Wildlife in a Liberal Education." *Transactions of the Seventh North American Wildlife Conference*, 8–10 April 1942: 485–89.

———. 1943. "Deer Irruptions." *Transactions of the Wisconsin Academy of Sciences, Arts, and Letters* 35:351–66.

———. 1944a. "Conservation: In Whole or In Part?" In Flader and Callicott 1991, 310–19.

———. 1944b. "The Land Health Concept and Conservation." In Callicott and Freyfogle 1999, 218–26.

———. 1944c. "Land-Health in S. W. Wisconsin." *The Aldo Leopold Archives*, http://digital.library.wisc.edu/1711.dl/AldoLeopold, Lectures 1933–1947: 868–70.

———. 1945. "Review of Young and Goldman, *The Wolves of North America*." In Flader and Callicott 1991, 320–22.

———. 1946a. "The Deer Dilemma." *Wisconsin Conservation Bulletin* 11 (8–9): 3–5.

———. 1946b. "The Land-Health Concept and Conservation." In Meine 2013a, 512–18.

———. 1947a. "Mortgaging the Future Deer Herd." *Wisconsin Conservation Bulletin* 12 (9): 3.

———. 1947b. "Original Foreword to *A Sand County Almanac*." In Meine 2013a, 873–78.

———. 1949. *A Sand County Almanac, and Sketches Here and There*. New York: Oxford University Press.

Long, Zachary T., John F. Bruno, and J. Emmett Duffy. 2011. "Food Chain Length and Omnivory Determine the Stability of a Marine Subtidal Food Web." *Journal of Animal Ecology* 80 (3): 586–94.

Luat-Hūʻeu, Kūpaʻa, K., Kawika B. Winter, Mehana B. Vaughan, Nicolai Barca, and Melissa R. Price. 2021. "Understanding the Co-Evolutionary Relationships between Indigenous Cultures and Non-Native Species Can Inform More Effective Approaches to Conservation: The Example of Pigs (Pua'a; Sus scrofa) in Hawai'i." *Pacific Conservation Biology* 27 (4): 442–50.

Lubchenco, Jane, and Bruce A. Menge. 1978. "Community Development and Persistence in a Low Rocky Intertidal Zone." *Ecological Monographs* 48 (1): 67–94.

Lyte, Mark. 2014. "Microbial Endocrinology and the Microbiota-Gut-Brain Axis." *Advances in Experimental Medicine and Biology* 817:3–24.

Marietta Jr., Don E. 1999. "Environmental Holism and Individuals." In *Environmental Ethics: Concepts, Policies, Theories*, edited by Joseph DesJardins, 238–46. Mountain View, CA: Mayfield.

McCann, Kevin, Alan Hastings, and Gary R. Huxel. 1998. "Weak Trophic Interactions and the Balance of Nature." *Nature* 395 (6704): 794–98. https://doi.org/10.1038/27427.

McShane, Katie. 2004. "Ecosystem health." *Environmental Ethics* 26 (3): 227–45.

Meine, Curt. 1987. "The Farmer as Conservationist: Aldo Leopold on Agriculture." *Journal of Soil and Water Conservation* 42 (3): 144–49.

———. 2004. "Emergence of an Idea." In *Correction Lines: Essays on Land, Leopold, and Conservation*, 117–31. Washington, DC: Island Press.

———. 2010. *Aldo Leopold: His Life and Work*. Madison: University of Wisconsin Press.

———, ed. 2013a. *Aldo Leopold: A Sand County Almanac & Other Writings on Ecology and Conservation*. Library of America Series No. 238. New York: Library of America.

———, ed. 2013b. "Chronology." In Meine 2013a, 849–57.

———. 2017. "Restoration and 'Novel Ecosystems': Priority or Paradox?" *Annals of the Missouri Botanical Garden* 102 (2): 217–26.

———. 2020. "From the Land to Socio-Ecological Systems: The Continuing Influence of Aldo Leopold." *Socio-Ecological Practice Research* 2 (1): 31–38.

———. 2022. "Land, Ethics, Justice, and Aldo Leopold." *Socio-Ecological Practice Research* 4:167–87.

Meine, Curt, and Gary P. Nabhan. 2014. "Historic Precedents to Collaborative Conservation in Working Landscapes: The Coon Valley 'Cooperative Conservation' Initiative, 1934." *Stitching the West Back Together: Conservation of Working Landscapes*, edited by S. Charnley, T. E. Sheridan, and G. P. Nabhan, 77–80. Chicago: University of Chicago Press.

Meynell, Letitia, and Clarisse Paron. 2021. *Applied Ethics Primer*. Halifax, Nova Scotia: Dalhousie Libraries Digital Editions.

Miller, Fred. 2017. "Aristotle's Political Theory: Political Naturalism." In *The Stanford Encyclopedia of Philosophy* (Winter 2017 Edition), edited by Edward N. Zalta.

Millstein, Roberta L. 2009. "Populations as Individuals." *Biological Theory* 4 (3): 267–73.

———. 2010. "The Concepts of Population and Metapopulation in Evolutionary Biology and Ecology." In *Evolution Since Darwin: The First 150 Years*, edited by M. A. Bell, D. J. Futuyma, W. F. Eanes, and J. S. Levinton, 61–86. Sunderland, MA: Sinauer.

———. 2013. "Natural Selection and Causal Productivity." In *Mechanism and Causality in Biology and Economics*, edited by Hsiang-Ke Chao, Szu-Ting Chen, and Roberta L. Millstein, 147–63. Dordrecht: Springer.

———. 2015. "Re-examining the Darwinian Basis for Aldo Leopold's Land Ethic." *Ethics, Policy & Environment* 18:301–17.

———. 2018. "Is Aldo Leopold's 'Land Community' an Individual?" In *Individuation, Process, and Scientific Practices*, edited by O. Bueno, R. Chen, and M. B. Fagan, 279–302. Oxford: Oxford University Press.

———. 2019. "Types of Experiments and Causal Process Tracing: What Happened on the Kaibab Plateau in the 1920s." *Studies in History and Philosophy of Science* 78:98–104.

———. 2020a. "Defending a Leopoldian Basis for Biodiversity: A Response to Newman, Varner, and Linquist." *Biology & Philosophy* 35 (12).

———. 2020b. "Functions and Functioning in Aldo Leopold's Land Ethic and in Ecology." *Philosophy of Science* 87:1107–18.

Mitman, Gregg. 2005. "In Search Of Health: Landscape and Disease in American Environmental History." *Environmental History* 10 (2): 184–210.

Molles, Manuel C. 2015. *Ecology: Concepts and Applications*. Vol. 7. London: McGraw-Hill.

Nelson, Michael P. 1996. "Holists and Fascists and Paper Tigers . . . Oh My!" *Ethics and the Environment* 1 (2): 103–17.

Newman, Jonathan A., Gary Varner, and Stefan Linquist. 2017. *Defending Biodiversity: Environmental Science and Ethics*. Cambridge: Cambridge University Press.

Nicholson, Teri E., Karl A. Mayer, Michelle M. Staedler, Jessica A. Fujii, Michael J. Murray, Andrew B. Johnson, M. Tim Tinker, and Kyle S. Van Houtan. 2018. "Gaps in Kelp Cover May Threaten the Recovery of California Sea Otters." *Ecography* 41:1751–62.

Nolt, John. 2006. "The Move from Good to Ought in Environmental Ethics." *Environmental Ethics* 28:355–74.

Norton, Bryan. 1984. "Environmental Ethics and Weak Anthropocentrism." *Environmental Ethics* 6 (2): 131–48.

———. 1988. "The Constancy of Leopold's Land Ethic." *Conservation Biology* 2 (1): 93–102.

———. 1992. "Sustainability, Human Welfare, and Ecosystem Health." *Environmental Values* 1 (2): 97–111.

———. 2002. "Epistemology and Environmental Values." In *Land, Value, Community: Callicott and Environmental Philosophy*, edited by Wayne Ouderkirk and Jim Hill, 123–32. Albany: State University of New York Press.

———. 2005. *Sustainability: A Philosophy of Adaptive Ecosystem Management*. Chicago: University of Chicago Press.

Odenbaugh, Jay. 2007. "Seeing the Forest and the Trees: Realism about Communities and Ecosystems." *Philosophy of Science* 74 (5): 628–41.

———. 2010. "On the Very Idea of an Ecosystem." In *New Waves in Metaphysics*, edited by Allan Hazlett, 240–58. New Waves in Philosophy Series. London: Palgrave Macmillan.

Odum, Eugene P. 1971. *Fundamentals of Ecology*. 3rd ed. Philadelphia: W. B. Saunders.

O'Neill, Robert V. 2001. "Is It Time to Bury the Ecosystem Concept (with Full Military Honors, of Course!)." *Ecology* 82 (12): 3275–84.

Ostrom, Elinor. 2009. "A General Framework For Analyzing Sustainability of Social-Ecological Systems." *Science* 325 (5939): 419–22.

Ouderkirk, Wayne. 2002. "Introduction: Callicott and Environmental Philosophy." In *Land, Value, Community: Callicott and Environmental Philosophy*, edited by Wayne Ouderkirk and Jim Hill, 1–18. Albany: State University of New York Press.

Paine, R. T. 1969. "A Note on Trophic Complexity and Community Stability." *American Naturalist* 103 (929): 91–93.

Papanikolaou, A. D., I. Kühn, M. Frenzel, M. Kuhlmann, P. Poschlod, S. G. Potts, S. P. M. Roberts, and O. Schweiger. 2017. "Wild Bee and Floral Diversity Co-Vary in Response to the Direct and Indirect Impacts of Land Use." *Ecosphere* 8 (11): e02008.

Peck, Steven L. 2009. "Whose Boundary? An Individual Species Perspectival Approach to Borders." *Biological Theory* 4 (3): 274–79.

Peck, Steven L., and Andrew Heiss. 2020. "Can Constraint Closure Provide a Generalized Understanding of Community Dynamics in Ecosystems?" https://www.biorxiv.org/content/10.1101/2020.01.28.924001v1.

Pershing, Andrew J., Michael A. Alexander, Christina M. Hernandez, Lisa A. Kerr, Arnault Le Bris, Katherine E. Mills, Janet A. Nye, Nicholas R. Record, Hillary A. Scannell, and James D. Scott. 2015. "Slow Adaptation in the Face of Rapid Warming Leads to Collapse of the Gulf of Maine Cod Fishery." *Science* 350 (6262): 809–12.

Post, David M., Martin W. Doyle, John L. Sabo, and Jacques C. Finlay. 2007. "The Problem of Boundaries in Defining Ecosystems: A Potential Landmine for Uniting Geomorphology and Ecology." *Geomorphology* 89 (1): 111–26.

Post, D. M., J. P. Taylor, J. F. Kitchell, M. H. Olson, D. E. Schindler, and B. R. Herwig. 1998. "The Role of Migratory Waterfowl as Nutrient Vectors in a Managed Wetland." *Conservation Biology* 12 (4): 910–20.

Proulx, Stephen R., Daniel EL Promislow, and Patrick C. Phillips. "Network Thinking in Ecology and Evolution." *Trends in Ecology & Evolution* 20 (6): 345–53.

Raine, André F., Scott Driskill, Megan Vynne, Derek Harvey, and Kyle Pias. 2020. "Managing the Effects of Introduced Predators on Hawaiian Endangered Seabirds." *Journal of Wildlife Management* 84 (3): 425–35.

Raine, Nigel E. 2018. "Pesticide Affects Social Behavior of Bees." *Science* 362 (6415): 643–44.

Rapacciuolo, Giovanni, Sean Maher, Adam Schneider, Talisin Hammond, Meredith Jabis, Rachel Walsh, Kelly Iknayan, Genevieve Walden, Meagan

REFERENCES

Oldfather, David Ackerly, and Steven Beissinger. 2014. "Beyond a Warming Fingerprint: Individualistic Biogeographic Responses to Heterogeneous Climate Change in California." *Global Change Biology* 20. https://doi.org/10.1111/gcb.12638.

Regan, Tom. 1983. *The Case for Animal Rights*. Berkeley: University of California Press.

Ricklefs, Robert E. 2008. "Disintegration of the Ecological Community: American Society of Naturalists Sewall Wright Award Winner Address." *American Naturalist* 172 (6): 741–50.

Ripple, William J., and Robert L. Beschta. 2005. "Linking Wolves and Plants: Aldo Leopold on Trophic Cascades." *BioScience* 55 (7): 613–21.

———. 2012. "Trophic Cascades in Yellowstone: The First 15 Years after Wolf Reintroduction." *Biological Conservation* 145 (1): 205–13.

Rohlf, Daniel, and Douglas L. Honnold. 1988. "Managing the Balances of Nature: The Legal Framework of Wilderness Management." *Ecology Law Quarterly* 15 (2): 249–80.

Román-Palacios, Cristian, and John J. Wiens. 2020. "Recent Responses to Climate Change Reveal the Drivers of Species Extinction and Survival." *Proceedings of the National Academy of Sciences* 117 (8): 4211–17.

Rosindell, James, Stephen P. Hubbell, and Rampal S. Etienne. 2011. "The Unified Neutral Theory of Biodiversity and Biogeography at Age Ten." *Trends in Ecology & Evolution* 26 (7): 340–48. https://doi.org/https://doi.org/10.1016/j.tree.2011.03.024.

Russow, Lily. M. 1981. "Why Do Species Matter?" *Environmental Ethics* 3 (2): 101–12.

SAGE (Sustainable Agriculture Education), and EPS (Economic & Planning Systems). 2007. "Martial Cottle Park Master Plan Case Study Report." https://parks.sccgov.org/sites/g/files/exjcpb961/files/CaseStudies-FINAL-20071029.pdf.

Sandler, Ronald. 2012. "Intrinsic Value, Ecology, and Conservation." *Nature Education Knowledge* 3 (10): 4.

Schulze, Ernst-Detlef, Erwin Beck, and Klaus Müller-Hohenstein. 2005. *Plant Ecology*. Berlin: Springer.

Shilling, Dan. 2009. "Aldo Leopold Listens to the Southwest." *Journal of the Southwest* 51 (3): 317–50.

Shin, Yunne-Jai, Guy F. Midgley, Emma R. M. Archer, Almut Arneth, David K. A. Barnes, Lena Chan, Shizuka Hashimoto, Ove Hoegh-Guldberg, Gregory Insarov, and Paul Leadley. 2022. "Actions to Halt Biodiversity Loss Generally Benefit the Climate." *Global Change Biology* 28 (9): 2846–74.

Shrader-Frechette, Kristin. 1996. "Individualism, Holism, and Environmental Ethics." *Ethics and the Environment* 1:55–69.

———. 1997. "Ecological Risk Assessment and Ecosystem Health: Fallacies and Solutions." *Ecosystem Health* 3 (2): 73–81.

Simberloff, Daniel. 2012. "Integrity, Stability, and Beauty: Aldo Leopold's Evolving View of Nonnative Species." *Environmental History* 17 (3): 487–511.

Simon, Herbert A. 2002. "Near Decomposability and the Speed of Evolution." *Industrial and Corporate Change* 11 (3): 587–99.

Singer, P. 1979. "Not for Humans Only: The Place of Nonhumans in Environmental Issues." In *Ethics and Problems of the 21st Century*, edited by K. E. Goodpaster, and K. M. Sayre, 191–206. Notre Dame, IN: University of Notre Dame Press, 191–206.

Smith, Subrena E. 2017. "Organisms as Persisters." *Philosophy, Theory, and Practice in Biology* 9 (14): 1–16.

Sommer, Ted, Bill Harrell, Matt Nobriga, Randall Brown, Peter Moyle, Wim Kimmerer, and Larry Schemel. 2001. "California's Yolo Bypass: Evidence That Flood Control Can Be Compatible with Fisheries, Wetlands, Wildlife, and Agriculture." *Fisheries* 26 (8): 6–16.

Soroye, Peter, Tim Newbold, and Jeremy Kerr. 2020. "Climate Change Contributes to Widespread Declines among Bumble Bees across Continents." *Science* 367 (6478): 685–88.

Sterelny, Kim. 2006. "Local Ecological Communities." *Philosophy of Science* 73:215–31.

Strauss, Sharon Y., and Rebecca E. Irwin. 2004. "Ecological and Evolutionary Consequences of Multispecies Plant-Animal Interactions." *Annual Review of Ecology, Evolution, and Systematics* 35:435–66.

Sullivan, Alexis P., Douglas W. Bird, and George H. Perry. 2017. "Human Behaviour as a Long-Term Ecological Driver of Non-Human Evolution." *Nature Ecology & Evolution* 1 (0065).

Sustainable Conservation. 2014. "Full Belly Farm Receives 2014 California Leopold Conservation Award." https://suscon.org/blog/2014/11/full-belly-farm-receives-2014-california-leopold-conservation-award/.

Tansley, Arthur G. 1935. "The Use and Abuse of Vegetational Concepts and Terms." *Ecology* 16 (3): 284–307.

Taylor, Charles. 1985. *Philosophy and the Human Sciences: Philosophical Papers 2*. Cambridge: Cambridge University Press.

Taylor, Paul W. 1981. "The Ethics of Respect for Nature." *Environmental Ethics* 3 (3): 197–218.

Thomas, Jeremy A., D. J. Simcox, and Ralph T. Clarke. 2009. "Successful Conservation of a Threatened *Maculinea* Butterfly." *Science* 325 (5936): 80–83.

Thompson, Ross M., Ulrich Brose, Jennifer A. Dunne, Robert O. Hall Jr., Sally Hladyz, Roger L. Kitching, Neo D. Martinez, Heidi Rantala, Tamara N. Romanuk, and Daniel B. Stouffer. 2012. "Food Webs: Reconciling the Structure and Function of Biodiversity." *Trends in Ecology & Evolution* 27 (12): 689–97.

Tscharntke, Teja, Ingo Grass, Thomas C. Wanger, Catrin Westphal, and Péter Batáry. 2021. "Beyond Organic Farming—Harnessing Biodiversity-Friendly Landscapes." *Trends in Ecology & Evolution* 36 (10): 919–30.

Vainshtein, Annie. 2022. "Foster City's Plan to Kill More Than 100 Geese, Potentially by Spinal Dislocation, Is Moving Forward." *San Francisco Chronicle*, July 20, 2022. https://www.sfchronicle.com/bayarea/article/Foster-City-s-plan-to-kill-more-than-100-geese-17318482.php.

Valiente-Banuet, Alfonso, Marcelo A. Aizen, Julio M. Alcántara, Juan Arroyo, Andrea Cocucci, Mauro Galetti, María B. García, Daniel García, José M. Gómez, and Pedro Jordano. 2015. "Beyond Species Loss: The Extinction of Ecological Interactions in a Changing World." *Functional Ecology* 29 (3): 299–307.

Valiente-Banuet, Alfonso, Elia Ramírez, Miguel Verdú, and Pedro Miramontes. 2014. "Plant Community Ecology." In *Frontiers in Ecology, Evolution and Complexity*, edited by Mariana Benítez, Octavio Miramontes, and Alfonso Valiente-Banuet. Mexico City: CopIt ArXives.

Van Auken, Paul. 2020. "Toward a Fusion of Two Lines of Thought: Creating Convergence between Aldo Leopold and Sociology Through the Community Concept." *Socio-Ecological Practice Research* 2 (1): 39–61.

Van Dyke, Fred. 2008. *Conservation Biology: Foundations, Concepts, Applications*. Cham, Switzerland: Springer.

Varner, Gary. 2020. "Response to Millstein." *Biology & Philosophy* 35 (15).

Vaughan, Mace, Matthew Shepherd, Claire Kremen, and Scott Hoffman Black. 2007. "Farming for Bees: Guidelines for Providing Native Bee Habitat on Farms." Portland, OR: The Xerces Society for Invertebrate Conservation, 1–44. http://www.pollinatorsnativeplants.com/uploads/1/3/9/1/13913231/farmingforbees.pdf.

Vucetich, John A., Jeremy T. Bruskotter, and Michael Paul Nelson. 2015. "Evaluating Whether Nature's Intrinsic Value Is an Axiom of or Anathema to Conservation." *Conservation Biology* 29 (2): 321–32. https://doi.org/10.1111/cobi.12464.

Wakefield, Jerome C. 1992. "The Concept Of Mental Disorder: On the Boundary between Biological Facts and Social Values." *American Psychologist* 47 (3): 373.

Warren, Julianne Lutz. 2013. *Aldo Leopold's Odyssey: Rediscovering the Author of "A Sand County Almanac."* Washington, DC: Island Press.

———. 2016. *Aldo Leopold's Odyssey: Rediscovering the Author of "A Sand County Almanac."* 10th Anniversary Edition. Washington, DC: Island Press.

Watson, James E. M., Hedley S. Grantham, Kerrie A. Wilson, and Hugh P. Possingham. 2011. "Systematic Conservation Planning: Past, Present and Future." *Conservation Biogeography* 1:136–60.

Wehr, Nathaniel H., Kealohanuiopuna M. Kinney, Nhu H. Nguyen, Christian P. Giardina, and Creighton M. Litton. 2019. "Changes in Soil Bacterial Community Diversity Following the Removal of Invasive Feral Pigs from a Hawaiian Tropical Montane Wet Forest." *Scientific Reports* 9 (1): 1–9.

Westra, Laura. 1997. "Why Norton's Approach Is Insufficient for Environmental Ethics." *Environmental Ethics* 19 (3): 279–97.

———. 2001. "From Aldo Leopold to the Wildlands Project: The Ethics of Integrity." *Environmental Ethics* 23:261–74.

Whitbeck, Caroline. 1978. "Four Basic Concepts of Medical Science." *PSA: Proceedings of the Biennial Meeting of the Philosophy of Science Association* (December).

Whittaker, Robert J. 1999. *Island Biogeography: Ecology, Evolution and Conservation*. Oxford, UK: Oxford University Press.

Whyte, Kyle. 2015. "How Similar Are Indigenous North American and Leopoldian Environmental Ethics?" Available at SSRN 2022038.

Young, Christian C. 2002. *In the Absence of Predators: Conservation and Controversy on the Kaibab Plateau*. Lincoln: University of Nebraska Press.

Zhao, Zhi-yuan, Yuan Xu, Lin Yuan, Wei Li, Xiao-jing Zhu, and Li-quan Zhang. 2020. "Emergency Control of *Spartina alterniflora* Re-Invasion with a Chemical Method in Chongming Dongtan, China." *Water Science and Engineering* 13 (1): 24–33.

Index

Figures and tables are indicated by f and t following the page number.

abiotic components of communities, 30–31, 33, 39–49, 52–58, 64, 68–69, 72

agriculture: biodiversity promoted by, 150; "biotic farming," 150n9; Coon Valley Erosion Project and, 136–37; exploitative forms of, 35, 83, 150; interdependence and, 33, 42, 114; land ethic's guidance for, 9, 17, 136, 158; Leopold's career implementing practices for, 17, 136; migratory geese and, 53, 65, 72–73; Sauk County farming community and, 113–14; seasonal, 137; soil fertility and, 83, 88, 107–8, 136

arguing for land ethic: competition and cooperation and, 116; conservation biology and, 120, 132; consistency argument, 115–19, 122, 130, 133–34; defense of premises for, 122–34; economic value and, 131–33; ethical basis of land ethic premise and, 125–26; extent of interdependence premise and, 122–23; food chains and, 117; historical and ecological evidence for, 117–18; holistic argument, 120–22; human ethics and, 115–19; interdependence and, 114–18, 122–32; intrinsic value and, 131–34; invasive species and, 121n11; kinds of interdependence premise and, 126–30; land communities and, 123–25; land health and, 118, 125; Leopold's communities and, 112–15; limits of human knowledge and, 121–22; love, respect, and admiration for land and, 119–20; main argument, 115–19; matter/energy flow and, 124; moral considerability of land communities premise and, 123–25; moral interests and, 123–25; moral obligations and, 119; necessities of completeness premise and, 130–31; organism view of communities and, 120, 130; other arguments, 119–22; overview of, 24–25, 112–15, 134; self-renewal and, 118; social/political value of human communities and, 129

argument from consistency, 115–19, 122, 130, 133–34

Aristotle, 125–29

ASCA. See *Sand County Almanac, A*

INDEX

Ashland Forest Resiliency Project, 146–49

"Battle for the Soul of Conservation Science, The" (Kloor), 9
Berkes, Fikret, 94
Beschta, Robert, 40n6, 106
Binkley, Dan, 29, 48
biodiversity: agricultural promotion of, 150; contemporary debates on, 95–100; definition of, 87; depletion of, 9; environmental policy and, 150–54, 158; land health and, 86–90, 93, 95–100, 106; Leopold on, 88n9, 94–95; shared aspects of views of, 97; soil fertility and, 107–8, 118
biological wholes: boundaries of, 59, 62, 66, 70, 77; causal interactions and, 62; congruence of, 70; continuity of, 70; as having beginnings and endings, 70, 73; land communities as, 55–56; organism view of communities and, 55; parts of, 66
biology, conservation. See conservation biology
biotic communities: abiotic included in, 54; biotic pyramid and, 14; as combination of ecosystem and community, 14; definition of, 2, 12–13, 54; as ecosystems, 13–15; environmental fascism and, 12, 16; myth surrounding, 10, 13–15; summary moral maxim and, 13, 54; terminological use of, 54. See also land communities
"biotic farming," 150n9
"Biotic Land Use" (Leopold), 94–95
"Biotic View of Land, A" (Leopold), 57, 91
Black Mountain land community, 66–68
Boorse, Christopher, 102
Bosque del Apache National Wildlife Refuge (BdANWR), 52–53, 73

boundaries. See community boundaries
Brando, Nicolás, 157
Brooker, Rob, 38

Callaghan, Terry, 38
Callicott, J. Baird: Darwin's influence on Leopold and, 19–20; ecosystem model and, 13–14, 61–63; ethics of Leopold and, 115n3; "holistic aspect" of land ethic and, 16n9; intrinsic value and, 132n18; land communities and, 54–55; land health and, 103–4; summary moral maxim and, 11n6
Caughley, Graeme, 29
causal interactions: biological wholes and, 62; interdependence and, 30, 34–46; land communities and, 62, 70–71; land health and, 87–90; positive and negative dimensions of, 34–41. See also ecological interactions
chains (food). See food chains
Chambers, Jeanne, 94
Chapin, F. S., III, 58
Clements, Frederic, 15, 23, 54, 56, 101
climate change, 9, 37, 44, 76, 107–9, 123, 143, 146–47, 153–54
communities. See biotic communities; community boundaries; ecological communities; land communities; organism view of communities
community boundaries: biological wholes and, 59, 62, 66, 70, 77; ecological communities and, 66–67; land communities and, 59–68, 74–75; matter/energy flow and, 58, 64–65, 68–72, 75–76; problem of, 59–64; responses to problem of, 64–68
competition, 20, 34–35, 68, 75, 91, 113, 116–17, 128, 139
conservation biology: arguing for land ethic and, 120, 132; eight Leopoldian subprinciples of, 139–43; human activity and, 9;

influence of Leopold on, 2, 6; Leopold on, 1, 51, 84, 108n24, 132, 135, 143n5, 146; organism view of communities and, 15; preservationist approach to, 9
"Conservationist in Mexico" (Leopold), 80
consistency argument, 115–19, 122, 130, 133–34
Convention on Biological Diversity, 154
Coon Valley Erosion Project, 136–39, 141, 143
cooperation, 40, 91, 116, 137, 139–40

Darwin, Charles: influence on Leopold, 8n4, 19–20; interdependence and, 20, 30, 57n3; land ethic and, 19–21; Leopold on, 8n4, 19–20; "proto-sociobiological" perspective myth and, 19–21
"Deer Irruptions" (Leopold), 28, 57
deLaplante, Kevin, 96–97
Descent of Man (Darwin), 19–20
diverse interests, 144, 146–49, 157–58
diverse values, 144, 150
diversity (biological). *See* biodiversity
Duffy, Linda, 146–47
Dussault, Antoine, 103

ecological communities: community boundaries and, 66–67; contemporary views of, 54–55, 58; land communities as incorporating elements of, 14, 23, 55–59, 68–69, 76, 96n16, 124; rehabilitation and, 138n2
ecological education, 142–43
ecological interactions, 42, 117, 123, 127–28, 155, 158. *See also* causal interactions
ecosystem health, 95n15, 101, 103, 110. *See also* land health
ecosystems: biotic communities as, 10, 13–15; human-free myth of, 17–18; land communities as incorporating elements of, 14, 23, 55–59, 68, 76, 96n16, 124; values and, 15–17
education (ecological), 142–43
Eliot, Christopher, 15, 35, 62–63, 71
Elton, Charles, 13, 30, 31, 57, 96
energy. *See* matter/energy flow
environmental ethics: contextual interpretative practice and, 12; human activity and, 100; influence of Leopold on, 2, 6, 115; organism view of communities and, 15; vulnerability as not much discussed in, 44n11. *See also* land ethic
environmental fascism, 11–12
environmental policy: balancing values and interests and, 143–49; biodiversity and, 150–54, 158; climate change and, 153; community conversations and, 146–47; concerns about subprinciples on, 156; Coon Valley Erosion Project and, 136–39, 141, 143; deploy variety of techniques subprinciple, 140; ecological education subprinciple, 142–43; ecological fascism and, 144; eight Leopoldian subprinciples of conservation policy, 139–43; feral pig depopulation and, 152–53; forestry and, 137, 140, 158; game management and, 141–43, 153; gather relevant scientific information subprinciple, 142–43; including and integrating all interests and values subprinciple, 139–40; interests and, 137–39, 143–49; land health and, 149–54; long view of conservation subprinciple, 141–42; matter/energy flow and, 158; moral obligations and, 140; other potential issues of concern on, 154–58; overview of, 136–39, 158–59; pluralistic approach to conservation policy and, 140; Principle of Integration of Land Uses and, 137–39, 147; recognizing

environmental policy (continued)
failures subprinciple, 140; resilience and, 138n2; rules for adjudicating competing interests and, 144; seeking cooperation principle, 139–40; soil fertility and, 141, 150–51, 154; systematic conservation planning approach contrasted with, 157; wilderness areas and, 154–56
ethic, land. See land ethic
ethical pluralism, 129, 140, 158
ethics (environmental). See environmental ethics

feral pig depopulation, 152–53
fertility, soil. See soil fertility
Flader, Susan, 3
Fleming, P. J. S., 106–7
food chains: arguing for land ethic and, 117; definition of, 91; human activity and, 17, 33, 117; interdependence and, 31, 33, 92–93, 98–100, 117; land health and, 24, 86, 90–92, 98–100, 105, 109–10, 118; long, 92–93, 98–100, 118, 154; matter/energy flow and, 91–92
food webs, 68–69, 100
forestry: environmental policy and, 137, 140, 158; land ethic's guidance for, 9; Leopold's career in, 4–6, 17, 31, 140–41; soil fertility and, 137
fountain of energy, 14, 56, 90–92
Freyfogle, Eric, 18
Full Belly Farm, 149–51

game management, 5–6, 28–31, 136, 141–43, 152–53. See also hunting
geese migration, 52–54, 57, 60–61, 69, 72–73
Ghiselin-Hull thesis, 55n2
Gleason, Henry, 15, 70–71
Goodpaster, Kenneth, 104–5, 124
Gounand, Isabelle, 74–75
Gray, Joe, 10
Gross, Lous, 58
Guha, Ramachandra, 17

Hastings, Alan, 58
health. See ecosystem health; land health
Helms, Douglas, 136, 138
Hettinger, Ned, 11
holism, 15–16, 40–41, 48, 120–22, 129–30
Holling, C. S., 12, 94
Honnold, Douglas, 17
Hooper, David, 100
Hubbell, Stephen, 71
human activity: conservation biology and, 9; ecosystems and, 17–18; environmental ethics and, 100; food chains and, 17, 33, 117; interdependence and, 33–34; land communities and, 73; land health and, 84, 87, 145; summary moral maxim and, 13
hunting, 6, 23, 48, 82, 141, 152–53. See also game management

integrity. See land health
interactions. See causal interactions; ecological interactions
interdependence: abiotic components included in, 30–31, 33, 39–49; agriculture and, 33, 42, 114; arguing for land ethic and, 114–18, 122–32; causal interactions and, 30, 34–46; central role in land ethic of, 14, 20, 23, 30, 76, 114; definition of, 31–41, 48; dependence as vulnerability and, 44–45; extent of, 122–23; food chains and, 31, 33, 92–93, 98–100, 117; human activity and, 33–34; interim account of, 41–46; Kaibab Plateau as example of, 28–29, 36–39, 48–49; keystone species and, 37; kinds of, 126–30; land communities and, 23–25, 30, 39–43, 48–49, 52, 57, 76; land pyramid and, 31, 33f, 35; Leopoldian understanding of, 31–41, 47–49; Leopold on, 23, 27, 31, 33, 48, 116–17, 125, 127n16; lines of dependency and, 14, 31–35, 32f,

57, 91, 131; moral obligations and, 25; as needed for completeness, 130–31; network of interactions and, 39n4, 45–49; objections to Leopoldian understanding of, 42–46; overview of, 23, 28–31, 49; population sizes and, 34, 36–41; positive and negative interactions and, 34–41; soil fertility and, 39; species interactions and, 24, 37, 69, 96, 98, 104, 109; trophic cascade and, 29; visualization of, 31, 32f, 33f; web of, 43, 46–47, 49, 98, 117–18; wolf-deer-plant, 28–31; wolves and, 28–31, 35–41, 48
interests. See moral interests
intrinsic value, 25, 54, 61, 131–34, 144
invasive species, 100, 108, 121n11, 143n6

Kaibab Plateau: causal interactions and, 37–39; criticism of Leopold's study of, 29; influence of Leopold's analysis of, 29; as interdependence example, 29, 36, 48–49; new attitude exemplified by study of, 28; positive and negative interactions and, 36–39; trophic cascade exemplified by, 29
Katz, Eric, 6
Kimmerer, Robin, 7, 46n13, 126n14
King, Martin Luther, Jr., 126
Kitchell, James, 53, 65, 72

Lackey, Robert, 103
land communities: abiotic included in, 52–58, 64, 68–69, 72; applications of, 72–76; arguing for land ethic and, 123–25; BdANWR and, 52–53, 73; as biological wholes, 55–56; Black Mountain land community, 66–68; causal interactions and, 62, 70–71; central role in land ethic of, 54–56; challenges of defining, 54–55, 58–64; climate change and, 76; community boundaries and, 59–68, 74–75; congruence and, 60–61, 65–68; contemporary ecology and, 54–58; criticism of, 54–56, 71; definition of, 2, 54–55, 58–64, 68; discussion of, 72–76; ecological communities and ecosystems, as incorporating elements of, 14, 23, 55–59, 68–69, 76, 96n16, 124; food webs and, 68–69; as forming a circuit, 47; human activity and, 73; interdependence and, 23–25, 30, 39–43, 48–49, 52, 57, 76; land pyramid and, 33f, 57; Leopold on, 47, 51, 56–57, 94–95; matter/energy flow and, 21, 24, 57–58, 64–66, 68–72, 75–76; meta–land communities and, 75–76; migratory geese and, 52–54, 60–61, 72–73; moral interests and, 61–63, 123–25; moral obligations and, 63, 131; open systems and, 59–61, 64, 69, 75; organism view of communities and, 15, 55–56; overview of, 23–24, 52–54, 76–77; parts and whole in, 66–67; persistence through time of, 69–71; Pond 18d and, 54, 60, 65, 70, 72; problem of boundaries for, 59–64; proposal for understanding, 68–72; responses to boundary problem for, 64–68; survival-relevant interactions and, 68–71; sustainability and, 58, 63–64, 70; terminological use of, 54; values and, 25, 54, 61, 131–34, 144; well-bounded systems and, 69
land ethic: agricultural guidance of, 9, 17, 136, 158; biotic community as ecosystem myth and, 10, 13–15; community and, 14–15; Darwin and, 19–21; debunking myths about, 10–21; definition of, 2–3, 114–15; ecological fascism and, 16–17; ecosystems as only entities of value myth and, 15–17; human-free ecosystems myth and, 17–18; inspiration for, 3–7; interdependence's role in, 14, 20, 23, 30, 76, 114; land communities' role in, 54–56;

land ethic (continued)
land health's role in, 11–12, 18–19, 82–83; Leopold on, 111, 114, 116, 118–19; moral obligations and, 14, 25, 63, 131; overview of, 1–7, 21–26; as pluralistic, 158; "protosociobiological" perspective myth and, 19–21; rejecting six myths on, 21; resilience and, 12; stability and, 11–12, 18–19; summary moral maxim and, 11–13, 16, 18, 114; wilderness areas' relation to, 17. *See also* arguing for land ethic

"Land Ethic, The" (Leopold): biotic community in, 54; context of, 12; definition of terms in, 12–13; ecological education in, 142; ecosystem interpretation of biotic community in, 14; history of ethics as history of "accretions" in, 16; human role in land communities in, 73; influence of, 6; integrity in, 85; interdependence in, 130; intrinsic value in, 131; land communities in, 57; land health in, 24; land pyramid in, 35; lines of dependency in, 35; obligation toward land and, 17; stability in, 84; writing of, 1

land health: addressing potential concerns with, 100–104; arguing for land ethic and, 118, 125; balance of nature view rejected and, 84; biodiversity's role in, 86–90, 93, 95–100, 106; causal interactions and, 87–90; causes of land sickness and, 87–90; central role in land ethic of, 11–12, 18–19, 82–83; circularity concerns about, 88–89; climate change and, 109; contemporary biodiversity-stability debates and, 95–100, 99t; definition of, 24, 93; environmental policy and, 149–54; feedback loops and, 89; food chains and, 24, 86, 90–92, 98–100, 105, 109–10, 118; game management and, 106–7; human activity and, 84, 87, 145; integrity and, 85, 94; invasive species and, 108; land pyramid and, 33f, 90–92; Leopold on, 40, 79, 82, 84, 90, 92, 94, 102; matter/energy flow and, 90–92; moral interests and, 25, 105; natural experiments and, 81–82; normativism and, 104; organism view of communities and, 83, 86, 101–2; overview of, 24, 80–85, 109–10; philosophical and scientific significance of, 104–9; as positive conception, 102; proposal for understanding Leopoldian conception of, 93–95; resilience and, 19, 94–95; self-renewal and, 19, 24, 40, 82–83, 86, 90–93, 101, 103, 109, 114, 118, 151; sickness in southwestern US and, 80–85; soil fertility and, 86, 88, 93, 107–8; stability used interchangeably with, 19, 83–84; summary moral maxim and, 84–85; sustainability and, 92–94; symptoms of land sickness and, 86–87; uncovering Leopold's views about, 85–93; wilderness areas and, 87; wolves and, 81, 89

"Land-Health Concept and Conservation, The" (Leopold), 85
"Land Pathology" (Leopold), 90
land pyramid, 24, 31, 33, 33f, 35, 57, 90–93, 105, 110
Laundré, John, 41
Leopold, Aldo: agricultural practices implemented by, 17, 136; on biodiversity, 88n9, 94–95; birth and early experiences of, 4; communities of, 112–15; on community boundaries, 60n4; on conservation biology, 1, 51, 84, 108n24, 132, 135, 143n5, 146; on Coon Valley, 136–39; on cooperation, 116; on Darwin, 8n4; death of, 5, 83, 112–13; early

INDEX

writings of, 5–6; on economic value, 132–33; education and training of, 4; on ethics, 116, 125; on food chains, 33; forestry and game management career of, 4–6, 17, 31, 136, 140–42; on game management, 106n22, 141–42; on grazing, 82; hunting experiences of, 6, 23, 28; influence of, 2–7, 9–10, 31, 87, 100, 115; on integrity, 94; on interdependence, 23, 27, 31, 33, 48, 116–17, 125, 127n16; on intrinsic value, 131; on invasive species, 108; on joint interests of stability, productivity, and beauty, 150; on Kaibab Plateau, 48; on kinship with fellow-creatures, 8n4; on land communities, 47, 51, 56–57, 94–95; on land ecology, 130–31; on land ethic, 111, 114, 116, 118–19; on land health, 40, 79, 82, 84, 90, 92, 94, 102; marriage and family of, 4; on moral interests, 137, 140, 150; on need for love, respect, and admiration for land, 119; on new science of ecology, 46; on organism view of communities, 79, 120; overview of, 1–7; on politics and economics, 127n16; reinterpreting of, 7–10; on resilience and, 94–95; on respect for fellow-members, 16; on self-interest, 116; on self-renewal, 90; social/equity and, 7–9; on striving toward harmony, 138; on summary moral maxim, 11; unifying potential of, 9–10; on values, 120; on wilderness areas, 17–18, 79; Wisconsin Shack of, 5–6, 112–13. See also land ethic; and specific works
Leopold, Carl, 4, 79
Leopold, Clara Starker, 4
Leopold, Estella, 4–5, 112
lines of dependency, 14, 31–35, 32f, 57, 91, 131
Long, Zachary, 100
Luat-Hūʻeu, Kūpaʻa, 152–53

Marietta, Don, 144
matter/energy flow: arguing for land ethic and, 124; community boundaries and, 58, 64–65, 68–72, 75–76; environmental policy and, 158; food chains and, 91–92; fountain of energy and, 14, 56, 90–92; land communities and, 21, 24, 57–58, 64–66, 68–72, 75–76
McShane, Katie, 102
Meffe, Gary, 12
Meine, Curt: arc of Leopold's career and, 140; biography of Leopold by, 3, 112n1; death of Leopold and, 113; environmental fascism charge and, 16n10; environmental policy and, 138n2; Leopold's attitudes on race, social justice, and progress and, 8; wildland protection and, 156
Meynell, Letitia, 126
migratory geese, 52–54, 57, 60–61, 69, 72–73
moral considerability, 25, 62–63, 124
moral interests: arguing for land ethic and, 123–25; balancing values and, 143–49; diverse interests and, 144, 146–49, 157–58; environmental policy and, 137–39, 143–49; including all relevant, 139–40; interest-relative entities and, 61–63; land communities and, 61–63, 123–25; land health and, 25, 105; Leopold on, 137, 140, 150; rules for adjudicating competing, 144; values and, 143–49
moral maxim, summary. See summary moral maxim
moral obligations, 14, 25, 63, 119, 131, 140

network of interactions, 23, 39n4, 45–49, 66–67, 126, 147
Newman, Jonathan, 75
Nolt, John, 119n9
Norton, Bryan, 19n13, 115n3

INDEX

obligations, moral, 14, 25, 63, 119, 131, 140
Odenbaugh, Jay, 42, 62–63
Odum, Eugene, 30, 38, 96
"On a Monument to a Pigeon" (Leopold), 155n14
O'Neill, Robert, 58
On the Origin of Species (Darwin), 20
open systems, 59–61, 64, 69, 75
organism view of communities: arguing for land ethic and, 120, 130; biological wholes and, 55; conservation biology and, 15; environmental ethics and, 15; land communities and, 15, 55–56; land health and, 83, 86, 101–2; Leopold on, 79, 120
Ouderkirk, Wayne, 34–35

Paron, Clarisse, 126
Peck, Steven, 73–74
Picasso, Valentin, 96–97
policy implications. *See* environmental policy
Politics (Aristotle), 125–26, 129
Pond 18d, 54, 60, 65, 70, 72
Post, David, 59–61, 69, 72
Principle of Integration of Land Uses, 25, 137–39, 147–49, 154, 158

Ragan, Jim, 112–13, 127
Rapacciuolo, Giovanni, 71
Regan, Tom, 11, 119
reintroduction of wolves, 40–41, 106–7, 124
resilience: definition of, 94; environmental policy and, 138n2; land ethic and, 12; land health and, 19, 94–95; self-renewal and, 90; stability and, 12
Ricklefs, Robert, 71
Ripple, William, 106
Rohlf, Daniel, 17
Rosen, W. G., 87

Sand County Almanac, A (ASCA) (Leopold): foreword to, 80; influence of, 6–7; inspiration for, 1, 4–5; land health in, 80, 85–86, 100; publication of, 5, 112; self-renewal in, 85–86, 90; writing of, 1, 4–5
Sauk County farming community, 113–14, 116, 130
Schulze, Ernst-Detlef, 58
Schwartz, Charlie, 112
self-renewal, 19, 24, 40, 82–83, 86, 90–93, 101, 103, 109, 114, 118, 151
Shrader-Frechette, Kristin, 54–56, 103
Simon, Herbert, 69
Singer, P., 119
Smith, Subrena, 101
soil fertility: agriculture and, 83, 88, 107–8, 136; biodiversity and, 107–8, 118; environmental policy and, 141, 150–51, 154; forestry and, 137; interdependence and, 39; land health and, 86, 88, 93, 107–8
species interactions, 24, 37, 69, 96, 98, 104, 109
stability. *See* land health
Sterelny, Kim, 66–67
Sullivan, Alexis, 37
summary moral maxim: biotic communities and, 13, 54; definition of, 11; human activity and, 13; land ethic and, 11–13, 16, 18, 114; land health and, 84–85; Leopold on, 11; myth of, 11–13
survival-relevant interactions, 68–71
sustainability, 19, 58, 63–64, 70, 92–94, 105, 149

Tansley, Arthur, 15
Taylor, Charles, 125–26, 128
Taylor, Paul, 105n20
"Thinking like a Mountain" (Leopold), 28, 35
Thomas, Jeremy, 36
Throop, Bill, 11
Tilman, David, 96
trophic cascade, 29–30, 106–7

Valiente-Banuet, Alfonso, 47
values: arguing for land ethic and, 131–34; balancing of, 143–49;

diverse, 144, 150; economic, 131–33; ecosystems and, 15–17; including all relevant, 139–40; including and integrating interests and, 139–40; intrinsic, 25, 54, 61, 131–34, 144; land communities and, 25, 54, 61, 131–34, 144; Leopold on, 120; moral interests and, 25, 62–63, 124, 143–49
Van Dyke, Fred, 6
Varner, Gary, 129
vulnerability, 44–45, 113, 117, 126

Wakefield, Jerome, 102
Warren, Julianne, 3, 19, 84n3, 85n5
Watson, James, 157
web of interdependencies, 43, 46–47, 49, 98, 117–18
webs (food), 68–69, 100
Westra, Laura, 17

Whyte, Kyle, 8, 126n14
wilderness areas: BdANWR and, 52–53, 73; environmental policy and, 154–56, 158; land ethic's relation to, 17; land health and, 87; Leopold on, 17–18, 79; protection of, 17–18, 154–56; "unspoiled wilderness," 81
Wisconsin Conservation Commission, 141
wolves: interdependence and, 28–31, 35–41, 48; land health and, 81, 89; Leopold's game management of, 28; predator loss and, 29, 81, 89; reintroduction of, 40–41, 106–7, 124

Yellowstone National Park wolf restoration, 40–41, 106–7, 124, 153
Yolo Bypass, 147–49